Secrets in the Solar System

Gatekeepers on Earth

Andrew Johnson

Book details:

Secrets in the Solar System: Gatekeepers on Earth

by Andrew Johnson

ISBN-13: 978-1981117550

ISBN-10: 1981117555

BISAC: Science / Space Science

Cover Design by Andrew Johnson (and NASA/JPL)

- Front cover image and back cover image are portions of MRO image PSP_007230_2170 (See chapter 7)
- Small images on Back Cover, from left to right, are:
- From: ESA Image ID 215131 (See chapter 4),
- From: JPL/NASA Spirit Rover Image from Sol 1402 - 2P250825588EFFAW9DP2432R1M1 (See chapter 8)
- From: ESA Image of Phobos from 15 Mar 2010 (APOD) (See chapter 14)
- From: MRO Image PSP_009342_1725 (See chapter 15)

The End of Totality – 21 Aug 2017, South Carolina
Photo by Andrew Johnson

No Wikipedia links included! (Wikipedia censors important information in real-time – use it at your peril!)

"Space isn't remote at all. It's only an hour's drive away if your car could go straight upwards."

Sir Fred Hoyle (1915-2001), British astronomer.

Other Books by Andrew Johnson

These are available as free PDF downloads, as well as other formats:

9/11 Finding the Truth (ISBN: 978-1548827618)

What really happened on 9/11? What can the evidence tell us? Who is covering up the evidence, and why are they covering it up? This book attempts to give some answers to these questions and has been written by someone who has become deeply involved in research into what happened on 9/11. A study of the available evidence will challenge you and much of what you assumed to be true.

9/11 Holding the Truth (ISBN: 978-1979875981)

The truth about what happened to the World Trade Centre on 11 September 2001 was discovered by Dr Judy Wood, through careful research between 2001 and 2008. The author of this book, Andrew Johnson, had a "good view" of how later parts of Dr Wood's research "came together." Not only that, he was also involved in activities, correspondence and research which illustrated that this truth was being deliberately covered up. This book is a companion and follow-up volume to "9/11 Finding the Truth" – and documents ongoing (and successful) efforts to keep the truth out of the reach of most of the population. Evidence in this book, gathered over a period of 12 years, shows that the cover up is "micro-managed," internationally and even globally. The book names people who are involved in the cover up. It illustrates how they often stick to "talking points" and seem to have certain patterns of behaviour. It attempts to illustrate how difficult it is to prevent the truth from being marginalised, attacked and "muddled up."

Climate Change and Global Warming... Exposed: Hidden Evidence, Disguised Plans (ISBN: 978-1976209840)

This book collects together, for the first time anywhere, a range of diverse data which proves that the whole issue of "climate change" is more complicated and challenging than almost all researchers are willing to consider, examine, or entertain. For example, this book contains astronomical data which most climatologists will not discuss in full. Similarly, the book contains climate and weather data that astronomers will not discuss. The book contains some data that neither astronomers nor climatologists will discuss. It contains some data that no scientists will appropriately discuss. It will show the reader why the climate change/global warming scam was invented, and it will illustrate how the scam has been implemented.

Table of Contents

Dedication

I would like to dedicate this work to all those who are truly open to finding the truth. Those with an open heart and an open mind. to those who oppose deception and are willing to work to expand knowledge without feeling the need to serve vested interests. I dedicate this to those who will try to break through the control mechanisms and reveal the truth, to the best of their ability and knowledge.

Author Biography

 Andrew Johnson grew up in Yorkshire and graduated from Lancaster University in 1986 with a degree in Computer Science and Physics. He worked in Software Engineering and Software Development, for about 20 years. He has also worked full and part time in lecturing and tutoring and he now works for the Open University (part time) tutoring students, whilst occasionally working on various small software development projects.

He became interested in "alternative knowledge" in 2003, soon after discovering Dr Steven Greer's Disclosure Project. Andrew has given presentations and written and posted many articles on various websites about 9/11, Mars, Chemtrails and Anti-gravity research, whilst also challenging some of the authorities to address some of the data he and others have collected.

Andrew is married and has two children. You can email Andrew Johnson on ad.johnson@ntlworld.com.

His website is www.checktheevidence.com.

Acknowledgements

Thanks to everyone that has helped in developing this research and to those who make the information, herein, freely available.

Richard D Hall, Anthony Beckett, Douglas Gibson

Important Researchers Quoted:

Richard C Hoagland

Dr Mark Carlotto

Sir Charles Shults III

Alan D Moen

Ron Bennett

Efrain Palermo

Dr Gilbert Levin

Dr Brian O'Leary

Dr Tom Van Flandern

Anonymous, Germany

1. Preface

Why this Book Exists

From the age of about 8, I have been interested in Astronomy and space exploration. I was 4 years old when the alleged Apollo moon landings took place. I used to be a committee member in the Coventry and Warwickshire Astronomical Society (UK) in 1987 and 1988. I even gave a couple of presentations at Society Meetings. I have tried to keep up with most of the main developments in space exploration and related topics. However, when I gained access to "always on" broadband internet in 2003, I began to find out a lot more about the Solar System, especially as the results of ESA and NASA missions were free to download... In other research, which I have written up in my other books, I came to know how secrets are being kept – about many things. These secrets are of world-changing significance. Since I came to know this, I have tried in various ways, to communicate what I have learned. I hope you find this "communication" useful in your journey of discovering how the world really is...

Is There Anybody Out There...?

The conventional view of the Solar System is, on the whole, somewhat dull to the average person and for some, the unmanned Space Exploration Programme has been far more interesting and exciting than the manned space programme. In the last 50 years, those who have taken an interest in the Solar System have witnessed the discovery of new Planetary Ring Systems, many new moons orbiting Saturn and Jupiter, active volcanoes on Jupiter's moon Io, Neptune's dark spot, Aurorae in the polar regions of Jupiter and Saturn and hundreds of bewilderingly beautiful images from the Hubble Space Telescope. Perhaps one of the high points is Saturn's incredible ring system – and an array of moons – which have been studied in incredible detail by the Cassini Probe. There are many mysteries, such as Jupiter's Great Red Spot (which has been present for at least 300 years) and Saturn's rings themselves. For those people who move from the "science fiction worlds" portrayed in Star Trek, Star Wars and Babylon 5 to a study of our Solar System, however, unless they are excited by weird geology or extreme weather systems, there seems, according to mainstream thinking, to be little to "write home about".

Over the last few years, following a fair amount of research I have conducted, I have concluded that this "somewhat dull" view of the Solar System should not be held so strongly – by so many in the science community. In the data we have got back from space missions, some people have pointed out a number of prominent anomalies – both on the Moon and elsewhere in the Solar System. These anomalies might indicate that there may be more to our "local neighbourhood" than randomised rock structures and the exotic atmospheric chemistry, which are the "bread and butter" of those who study Astronomy and

Planetary Science. I now contend strongly that the pages of our Encyclopaedias and other reference books and "officially recognised" websites should include a more considered analysis of certain features in the Solar System, which have been photographed at sufficient resolution to show that they do not seem to fit into the "dull picture" that is so often promoted.

Some Rhetorical Questions You May Be Asking...

If these anomalies are significant, why haven't they been discussed and analysed in a more serious manner? Would anyone really be "keeping secrets" about these anomalies? Why has any discussion been left in the "doldrums" of the Daily Mail, or similar publications?

Here is a further list of questions that some people might ask themselves – or me - when seeing a book with a title like this...

- Why would anyone write a book like this one? Surely there are just too many books about UFO's, Aliens and all that stuff? Why would anyone question what the public and open science institutions and establishments show us and tell us? What on earth (or not on earth) would they wish to hide?

- Science – particularly astronomy – is free, open and merely concerned with the study of the cosmos. Why would there be any "secrets in the Solar System"?

- Come on, Andrew, there aren't any are there? If there was any important evidence regarding discoveries in space exploration, we would have been told by now, surely? Why would any such discoveries be covered up?

I'd like you to consider what many of these questions are concerned with, however. They are concerned with motivation. Imagine if all matters regarding enquiry – criminal/police investigations, medical investigations and legal matters had to establish a motive before any evidence was considered and one had to give a plausible motive before *any* investigation was started in earnest. People might also ask:

- Aren't you just writing this book to make money – or something?

- And anyway, you're not a professional Astronomer, are you?

Let's say I was writing this book to make money, would that change what is shown in it? If you were to get hold of a free cookbook or one for which you paid money, which one would you trust more? Or, if you got hold of a free maths text book and compared it to one that cost a little money, would the free one be less trustworthy or valid than the one you paid money for?

It is true, I am neither a professional astronomer nor am I a professional writer or author – all my works (so far) are self-published or web-published. So, does that make what I include in this book more "suspect" or "unreliable"? I hope you will check the sources of information I have included, then you will know if I am being truthful or not.

By reading this far, I do hope you have started to separate considerations about evidence from considerations about motives and other issues. Considering motives *before* basic evidence can cause free enquiry to be inhibited or even prevented. If you can separate "evidence" from "motivation" then you can evaluate each of the sections of this book on its "own terms."

Wrong Assumptions? Wrong Information?

As I have said in presentations I have given, and in the author biography, my main areas of expertise are in the field of information technology – particularly computing and programming - in the field of real-time software. Outside of this, I have completed a science degree and, I would say, I have a fairly sound understanding of the fundamentals of mainstream topics in Physics, Chemistry and Biology. In relation to this book, however, we are looking mainly at image artefacts to do with geology and possibly archaeology. I am not an expert in these fields, so I appeal, at the outset of this journey, to anyone that finds mistakes in this work, to please politely point them out to me, with references to any corrections that are needed. In my role as a tutor to students, this is what I am expected to do, so it would be churlish of me to disregard any corrections or valid criticisms of the research I present here. For what it is worth, I have been presenting (online) a fair proportion of what is in this book, in various forms for over 10 years, with little feedback or informed criticism. Others, more expert than I, have also already presented a fair proportion of this evidence, but not been given satisfactory answers or feedback to what has been shown.

A Wider Picture

I hope that, by the time you have read to the end of this book, you will be open to considering how perception of discoveries in space exploration has been managed. I will try to offer some clues as to why this perception management has been in operation, though you will probably be able to think of quite a few reasons for yourself. Why has discussion and analysis of these anomalies been suppressed and ridiculed?

2. Introduction

To set the scene for our "journey into space," I will briefly cover the history of space exploration.

Unmanned Space Missions

I have been interested in this subject since I was a child. I was born in 1964 – near the beginning of the so-called "space age." Looking back, it almost seems like a "golden age of space technology"…

The space race, according to "official sources," is now over 60 years old. The simple Russian Sputnik radio transmitter was launched on 04 October 1957[1] and it completed about 1400 orbits of the earth (each one took about 98 minutes[2]). This was followed only a few months later by the USA's explorer 1 satellite, which carried a Geiger-Mueller tube[3]. This satellite was the first one to detect the so-called Van Allen zones of radiation.

The Russian "Luna 1" or "Lunik 1" was the first probe to reach the moon, in early 1959[4], but this only contained about five instruments and no camera.

The first pictures of the moon, taken from an unmanned probe, came from Ranger 7 on 31 July 1964[5]. The pictures were taken before the probe crash-landed there.

The first "soft moon landing" took place in January 1966[6] – The Russian Luna/Lunik 9 probe was able to take a number of photos.

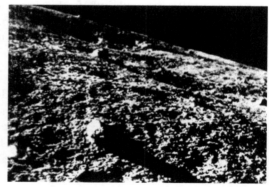

Lunik 9   First Image of Moon's Surface from Lunik 9

The Russians had already made an attempt to get to Mars in 1960 with Marsnik 1, but this failed.

Both the USA and the USSR continued to launch unmanned probes to the moon and Mars throughout the 1960s and 1970s – with varying levels of success. The Russians claimed their Lunakhod 1 and 2 rovers successfully drove on the surface of the Moon in 1970 and 1973, where they performed a number of experiments and took many photographs.

The first successful fly-by of Mars was made by the US Mariner 4 Probe, in July 1965[7] (an event covered in a radio broadcast by Richard C Hoagland[8], who we shall mention later).

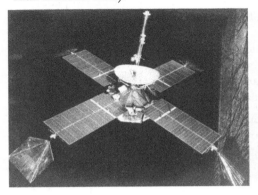

Mariner 4[9]

Mariner 4 image of Martian Surface, 1965[10]

Since the 1970s, unmanned probes have visited all planets in the Solar System – and a number of the moons of Jupiter and Saturn. Additionally, there have been several missions to Comets. The European Space Agency's mission "Giotto" flew into the coma of Halley's Comet in March 1986 and returned many photos and much additional data. Since then (in 2003), ESA has launched Mars Express, a sophisticated orbiter, which also carried the ill-fated Beagle 2 lander. In October 2016, ESA's Schiaparelli Mars Lander was "lost" just 1 minute before landing.[11]

In this book, I will mainly be focusing on data from several unmanned missions. The years listed below are the ones in which the respective probe began returning data.

- 1966-7 – Lunar Orbiter 2 - 5[12]
- 1976 - Viking 1 and 2 Mars Landers and Orbiters [13]
- 1995 – SOHO – Solar Observatory[14]
- 1998 - Mars Global Surveyor (abbreviated to MGS)[15]
- 2003 – Mars Express (an ESA mission)[16]
- 2004 - Cassini (Saturn, Iapetus)[17]
- 2004 - Spirit and Opportunity Rovers (Mars)[18]
- 2008 - Phoenix Lander (Mars)[19]
- 2008 – Mars Reconnaissance Orbiter (abbreviated to MRO)[20]
- 2009 - Lunar Reconnaissance Orbiter (LRO)[21]
- 2012 - Curiosity Rover (part of the Mars Science Laboratory)[22]

I will refer to other missions, too. These missions have mostly been designed and run by NASA.

In chapter 4, I will include a discussion of the Mars anomaly that most people (who are interested in these topics) will have heard of – the so-called "Face." We will also see there are several other features near the "Face" which are worthy of study. Much more information about the analysis of the Face can be found in Dr Mark Carlotto's book "The Cydonia Controversy"[23] and in Richard Hoagland's books "Monuments of Mars" and "Dark Mission". I will be referring to the Carlotto's and Hoagland's research quite frequently.

I will cover some of the anomalies found on the Moon in chapter 3 and refer to data from several Lunar Orbiter missions (but not Apollo missions).

Manned Space Missions

Although I will not really be discussing manned space missions in this book, I wanted to mention them briefly here for completeness.

The alleged first orbit of the earth by a person was made by Yuri Gagarin on 12 April 1961 (I say "alleged" because some people have found evidence that Gagarin was not in the capsule when it was aloft[24]. Similarly, it was revealed in 1999 that Gagarin did not actually land in his capsule[25].)

Following the alleged Gagarin mission, there was a period of increasingly "daring" manned missions and programmes by the USA (Mercury and Gemini primarily) and the USSR (Soyuz and Salyut).

Most authors would perhaps then talk about the alleged Apollo Moon landings between 1969 and 1972. At this point, I cannot accept, myself, that the Apollo mission record is truthful, accurate and consistent[26]. I hope to cover this in a separate book. What I have learned about the Apollo missions leads me to certain conclusions about what the programme was for. Conclusions about Apollo then become relevant to the claims made for the Mars rover missions, which I discuss in chapter 17.

What is the Purpose of Space Exploration?

The reasons for space exploration could be said to include one or more of the following:

- To find out more about the earth.

- To find out more about our evolution.

- To find out more about how we can use space-borne technologies to enhance our lives here on terra-firma.

- To determine if life (intelligent or otherwise) exists elsewhere in the Solar System or the galaxy.

In this book, I am most interested in considering this last reason and, as we shall see, some of the "mission statements" from those in our list specifically state they will be looking for life. One of the less-considered, but related, reasons to explore space might be:

- To determine if life (intelligent or otherwise) <u>has ever existed</u> elsewhere in the Solar System.

What would happen if scientists discovered irrefutable evidence of past or present extra-terrestrial life in the Solar System? Would they "tell us the truth, the whole truth and nothing but the truth" about such a discovery? Or, would the "scientific technological elite" mentioned in Eisenhower's final address to the USA[27], become the "gatekeepers" of "Secrets in the Solar System?"

3. Fly me to the Moon…

If you were only to view the Apollo mission footage and most of the data shown in text books and documentaries about the Moon, you would be most likely to conclude that the Moon is a lifeless world, because of its hostile environment – with any point on the surface flipping between searing heat and frigidity every 2 weeks. It is simply a world of rocks, dust and craters…

However, some interesting observations have been made, over the years. The observations and images discussed in this chapter may not be the result of the activities of extraterrestrial life – past or present – but I would argue that these features are not obviously the result of easily recognised or described geological processes either, hence their inclusion. At the very least, they may indicate that the moon is not a "hot-cold, dead rocky world…"

TLP and Lunar Anomalies

An interesting discussion of what "lunar matters" we might not be being told about can be found in Ingo Swann's book "Penetration[28]". Swann references a very interesting NASA "Technical Report" R-277 from 1966 – "Chronological Catalog of reported lunar events"[29]. (Among the listed authors is famous UK Astronomer Patrick Moore!) The Abstract of the report reads:

> *A catalog of reports of lunar events, or temporary changes on the moon, has been compiled based on literature covering more than four centuries. In most cases, the original reference has been consulted… Each entry includes a brief description and date of the observation, the name of the observer(s), where these are known, and the reference.*

The report lists what are known as "Transient Lunar Phenomena" – and these are still observed today[30]. Most of these phenomena are, by definition, short-lived – lasting a few hours or perhaps 2 days. They are only observed by a very small number of people, so they have even been dismissed by mainstream scientists and experts. They are rarely, if ever, discussed outside astronomical forums. The NASA catalogue lists 579 observations over the period of 1540 to 1967. Not surprisingly, some of these observations were made by very famous astronomers such as Giovanni Domenico Cassini (1625 - 1712) – who discovered the Saturnian Moon Iapetus (see chapter 18) and William Herschel (1738 – 1822) who discovered Uranus. They made the following observations:

Cassini

Date	Location	Description
1671 Nov 12	Pitatus	Small whitish cloud
1672 Feb 3	Mare Crisium	Nebulous appearance
1673 Oct 18	Pitatus	White Spot.

Herschel

1787 - Apr 19	Dark side	Three "volcanoes." The brightest, 3 '57 ".3 from N limb, the other two much farther toward the centre of the moon.
1787- Apr 20	Dark side	Brightest "volcano" even brighter and at least 3 mi in diameter.
1787 May 19-20	Aristarchus	Extraordinarily bright.

Lunar Orbiter Images

There are one or two other interesting photos from the Lunar Orbiter Project page[12] (see below). So, what was the Lunar Orbiter Project?

The Lunar Orbiter Program, managed by Langley Research Center, was one of three unmanned programs undertaken by NASA to collect data and help select site for the manned landings. The low orbits around the moon provided extensive photographic coverage of specified areas. The program was a series of five unmanned lunar orbiter missions from 1966 through 1967. Intended to help select Apollo landing sites by mapping the Moon's surface, they provided the first photographs from lunar orbit.

Lunar Orbiter 5, the last of the Lunar Orbiter series, was designed to take additional Apollo and Surveyor landing site photography and to take broad survey images of parts of the Moon's far side. The spacecraft acquired photographic data from August 6 to 18, 1967, and readout occurred until August 27, 1967. A total of 633 high resolution and 211 medium resolution frames at resolution down to 2 meters.

This shows frame 67-7530:

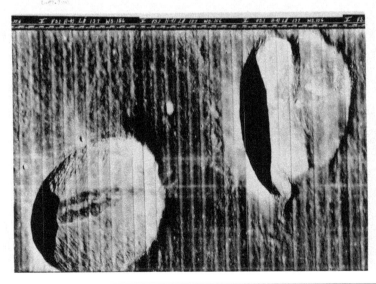

I have zoomed in on the lower left portion of the image and rotated it:

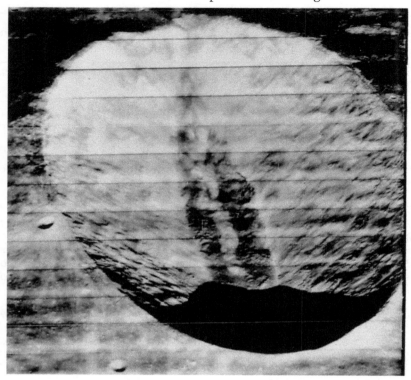

Further zooming reveals details on the crater floor:

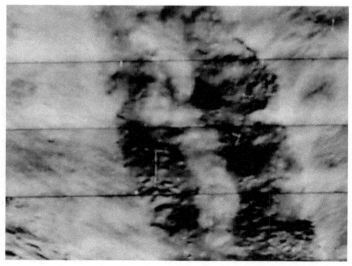

Is this simply an impact crater or a volcanic crater?

Another image I found on the Lunar Orbiter project page[31] appears to show some type of circular feature which either isn't a crater, or it is a crater which has somehow got filled in:

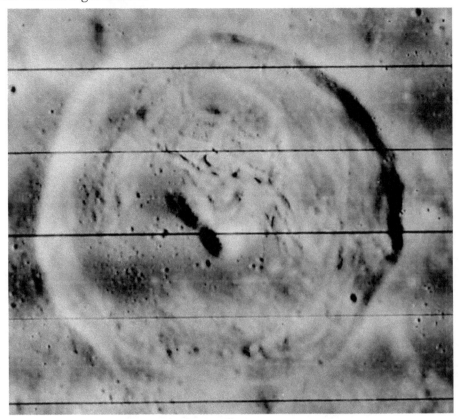

One can wonder what the unusual tracks are... but we rarely hear such questions asked and discussed on any astronomy programmes, websites or in presentations.

Aristarchus Crater

This crater is named after the ancient Greek Astronomer, Aristarchus. It is at Lat: 23.7°N, Long: 47.4°W and has a diameter of 40 km (A blog about TLP's on the Armagh Planetarium's Website mentions this crater and notes:[32]

An interesting fact that was discovered when all the 579 official TLPs were recorded was that a third of these were observed in the vicinity of Aristarchus crater on the north-west part of the Moon's near side. The feature itself is already quite prominent as it is already considered to be quite bright itself and easy to identify when it is highlighted by the occurrence of earthshine on the Moon. (This is when reflected sunlight from the Sun on the Earth lights up the dark portion of the Moon).

It then includes this image:

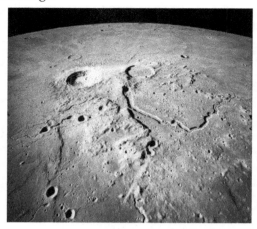

Aristarchus Crater (Moon). This image was allegedly taken from Apollo 15 (the curvature seems rather odd to my "ignorant" brain...)[33]

There is also an interesting post on an LRO website (LROC) which fanfares new high-resolution images of the crater[34].

LRO image of Aristarchus Crater

Above this image it says (my emphasis added):

*Aristarchus crater is located on the southeast edge of the Aristarchus Plateau. This gaping crater is 40 km wide and 3.5 km deep. **The ledges forming the wall of the crater, which <u>look a lot like those of a strip mine</u>, are actually blocks of pre-impact crustal and surficial rocks** that slumped into the crater during the late stages of its formation. The impact that formed this crater occurred on a mare-highland boundary and thus excavates a variety of rock types.*

The LRO page, unlike the Armagh Planetarium page, mentions nothing about TLP's. You can analyse the crater yourself, now, in high resolution with the LROC viewer[35]. Another interesting page about the Aristarchus Crater entitled "Hubble Space Telescope Looks at the Moon to Prospect for Resources"[36] holds this image (which is Hubble Telescope imagery overlaid on "simulated topography"). Again, there is no mention of TLP on this page, despite the very striking images.

A Hubble-telescope/3D Model view of Aristarchus[36]

NASA/Goddard Space Flight Centre Scientific Visualization Studio
Additional credit to Zoltan G. Levay (STScI)

"The Living Moon"

Additional images of the Aristarchus Crater can be found on John Lear's "The Living Moon" website.[37] Whilst I won't claim to have examined all the images held there, and I can't say I am convinced by the claims he makes for the images he shows, he does at least attempt to reference some of the material he uses. I still find some of it to be highly interesting and I do get the feeling we are not being told the whole truth about some of these images. Additionally, on the page I reference above, Lear accurately mentions the same information about TLP's that I included earlier.

Other websites and postings talk about the Aristarchus Crater – such as the "Above Top Secret" forum.[38] Some of the images they reference don't seem to be easily traceable, however.

Though we will be exploring more the idea that TLP might not be only due to geological or impact events, the LRO page is typical in that it does not attempt to "join up" observations of different types and come up with a suitable explanation. For example, why not relate back to Herschel's observations? Perhaps he was indeed seeing volcanic activity?

Other TLP may be caused by the impact of a large meteorite – a few tens of meters in diameter. (The Moon has no appreciable atmosphere to cause a meteorite to glow hot as it gets nearer to the surface.) We should not let the commonly held view of a static, unchanging, lifeless orb prevent us from considering there may be much, much more to the moon... Perhaps the closest we get to joining up observations of TLP with geological activity on a NASA website is an article entitled Evidence for Young Lunar Volcanism[39], in which we can see the following LRO image:

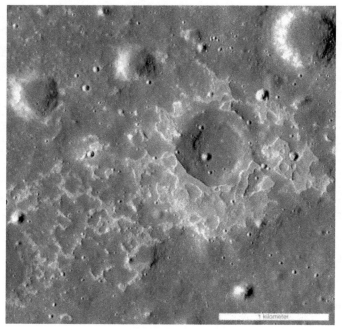

Evidence for more recent lunar vulcanism?

Who Built the Moon?

Is the moon a giant spaceship? Whilst I won't be going in to the detail given by Alan Butler and Christopher Knight in their book "Who Built the Moon?,"[40] this is certainly the sort of thinking that I like to indulge in, to see "where it takes me." Butler and Knight's book wasn't the first to consider this idea – earlier works by Don Wilson - Our Mysterious Spaceship Moon[41] and Secrets of Our Spaceship Moon[42] covered similar territory. That is, what if the moon *was* constructed and not formed during some kind of "cosmic calamity"? I have not really made a conclusion on the issue, as the evidence seems somewhat weak. Also, I struggle to comprehend that something the size of the moon would be a "spaceship" in the sense that it would or could be used to carry a large group of people from one world to another.

Is Someone Else on the Moon...?

Ingo Swann's "Penetration" book also references another book, from 1976. George H Leonard's Somebody Else is on the Moon[43], which includes pictures of what Leonard considers to be lunar anomalies. Though I found Swann's book very interesting, I cannot say the same for Leonard's book. This is because I cannot see what Leonard sees in the images in his book. I cannot see the indications of machinery and machining, which he claims. Leonard also treats the Apollo record as authentic (it is not) and this is repeatedly an issue in lunar anomaly research. Most of the images that Leonard references in his book seem to be difficult to find today – perhaps because back in 1976, the referencing system for the images was more basic. It's not clear how many of the images have been digitised and catalogued according to their original paper-based catalogue numbers. I did find some of the images Leonard used, but again, could not really see what Leonard could see. For example, plate 19 (given a reference of 67-H1179) in Leonard's book is a picture of the Tycho Crater[44] (he discusses it on pages 120 and 127 of his book). He highlights two features:

Tycho Crater, as highlighted by George Leonard

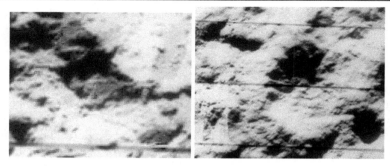

Arrowed Area is on the left, circled area is on the right.

In the circled area, I can see that it kind of appears like a "bite mark," but I don't see anything really weird about this – and as the photos are rather low in resolution, it seems we can't draw any firm conclusions about what we are seeing. It could just be natural rock formations (for the moon). Leonard labels plate 19 with "19. Coverings with glyph and manufactured object are indicated in Crater Tycho. (See pp. 120, 127.)"

Once we know that Tycho Crater is at Lat: 43.31°S, Long: 11.36°W, we can now look at Tycho on an LRO image on the interactive map. I have done this, and it is difficult to see the features that Leonard highlighted. The "bite" marks are actually in a crater called Tycho C. The alleged machine is not at all visible. I looked at an 8m/pixel image using the LRO map and found this in the area[45]:

Close up of the area **near** where Leonard said he could see a machine in the Lunar Orbiter Image[45] – I couldn't see any such machine, but this image is interesting because it looks like liquid was flowing there at one time...

Vitello Crater

This crater is located Lat: at 30.4°S, Long: 37.5°W and it has a diameter of 42 km, Depth: 1.7 km. Like the Ukert crater (covered later), it isn't perfectly circular, and it has some unusual channels in the bottom. Further, it has a strange conical feature in the 10 O'clock position.

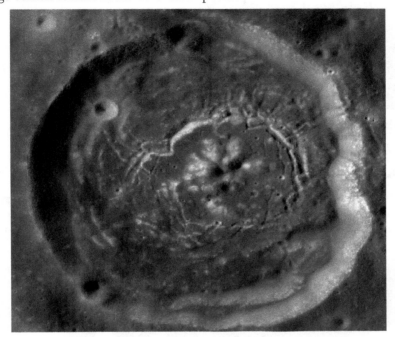

LROC Image of Vitello Crater

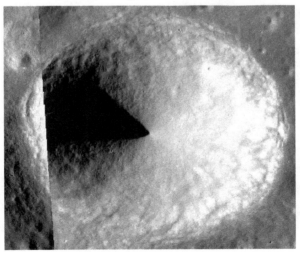

Closer image of conical feature in Vitello Crater[46].

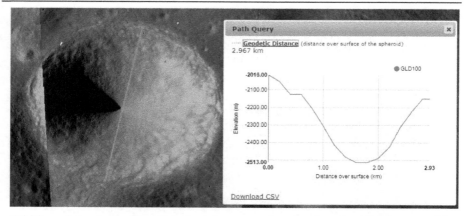

Width and depth profile of conical feature – showing it to be 3km across with a smoothly curving basin – down to a depth of about 500m.

Rock 'n' Roll

Returning to Leonard's book, an interesting image can be found in the form of Plate 14 (discussed on page 96 of his book. This image was quite difficult to track down on a NASA website[47].) It is actually from Lunar Orbiter 5 – frame h168-2[48]. It turns out that this is in the Vitello Crater!

Leonard labels this: *Two "rolling" objects on the floor: of Vitello Crater. Note that the smaller object "rolled" up, out of the crater.*

I don't think Leonard could possibly justify his explanation with such limited data to go on, but we can probably agree that the image looks unusual. The resolution of the 1960s photograph limits further identification. Leonard does himself "no favours" and makes it more likely that people will want to "debunk" his observations rather than try to examine them more closely...[49]

After much searching on the LRO QuickMap[50], we can see a higher resolution image which more clearly shows that it is a boulder – about 20m in width - at the end of a trail. It isn't a "monolith". It seems much lighter in colour than the surroundings.

Quite a number of these boulder trails have now been identified[51] in various Lunar Orbiter images. Below, some more boulder trails in the Vitello Crater are shown.

LROC Image - More boulder trails in the Vitello Crater[52]

It seems sensible to suggest that the trails' unusual appearance is due to both the aerial view we are seeing it from and the lower gravity on the moon.

The Blair Cuspids

We will now look at Plate 15 in Leonard's "Someone else is on the Moon." I think this is a rather more interesting image, which is reproduced below, with someone else's annotations.

I shouldn't need to point out the 4 bright points near the digits "1" and "7" which seem to be arranged in an L shape. Are these pillars or obelisks? Are they "architecture of some kind?"

The "Blair Cuspids"

An excellent summary of the story of their discovery is contained on a succinct page by Keith Fitzpatrick-Matthews[53].

> On 2 November 1966, NASA published a photograph of a region towards the western edge of the Sea of Tranquillity taken by Lunar Orbiter II (reference number LO2-61H3 [i]) at approximately 15.5°E 5.1°N. William Blair of the Boeing Institute of Biotechnology drew attention to a number of apparently anomalous features in the photograph, mainly a series of objects that cast long, clear shadows, contrasting them with the shorter shadows cast by objects that were more obviously boulders.

The article continues:

> A second photograph of the site was subsequently located by Lan Fleming (reference number LO2-62H3), taken 2.2 seconds after the first. This shows that the largest of the cuspids, number 5, is considerably broader at the base than was originally thought.

Matthews concludes, however, that their arrangement probably isn't significant because of probabilities. That is, there are so many boulders lying on the moon's surface that it is just possible an alignment may form naturally.

A very useful page on the "Astrosurf" website by Fran Ridge gives about ten links to different NASA images which show the objects and considers them worthy of study[54]. It lists several LRO image numbers which show the objects.

They are somewhat difficult to find on the Quick Map, but Fran Ridge's guide helps a great deal. The current QuickMap version isn't the highest resolution version available.

In the image (left), we can see some blurring and compression artefacts — perhaps due to a mixture of digital effects such as aliasing or mapping the image onto a flat surface etc.

2017 LRO QuickMap image of the "Blair Cuspids" (enlarged)[55]

The highest resolution image that can currently be found via a separate browser application. This browser application allows you to zoom and pan across the image[56] (see below).

Browser page for highest resolution image showing the Blair Cuspids.

The site of the Blair Cuspids can be found a little distance above the 2 craters in the middle of the zoomed-out browser image. I zoomed in to see the image below.

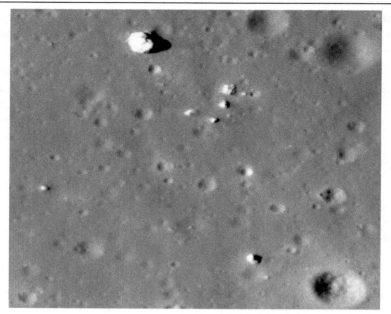

Highest resolution image showing the Blair Cuspids.

This image makes the shape of the objects a little bit clearer and we can also see that the "L" (or reversed "L") shape doesn't look quite as well-defined.

I have included another version of the image below with my own annotations and associated notes.

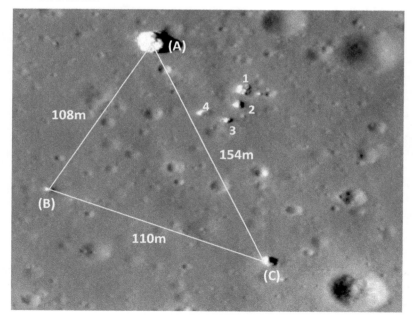

Using the LRO Map measurement tool, I found that there is perhaps an isosceles triangle in the arrangement of some of the stones A, B and C (within a margin of error). However, there are other boulders near these (if you view the whole scene) and if you were to draw lines between these, you may find other close alignments. (i.e. I would say I am in partial agreement with Keith Fitzpatrick -Matthews regarding this.) Boulder C is much bigger than B and C and very much looks like a boulder – not a tower or anything like that. So, let us have a closer look at objects 1-4.

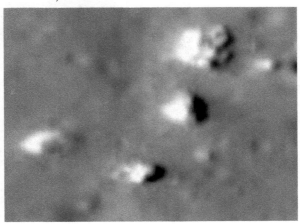

Object 1 looks like a squarish bolder with a square piece broken off it. Object 2 does seem to have a vaguely pyramidal shape, but the resolution is still not high enough to make any further deductions about its exact shape. Objects 3 and 4 are smaller and we can't really say much about their shape. Again, the alignment seems less exact, now that we are looking at higher resolution images.

All the objects 1-4 and A-C do look quite bright – their colour is much different from the lunar surface (regolith). However, we have already seen this in the earlier images of boulders.

For those wishing to study this image more closely, I recommend Mark Carlotto's extremely detailed report, published in 2002, called "3-D Analysis of the 'Blair Cuspids' and Surrounding Terrain."[57] This report includes some interesting additional observations in the form of 3D models he has created based on the way the scene is illuminated. Additionally, an apparent rectangular-shaped depression near the rocks is also studied. It would be interesting to see how much of this report is verified using the new images and data from LRO.

Ukert Crater

Another interesting feature I came across is the Ukert Crater, Lat: 7.8°N, Long: 1.4°E. It has a diameter of 23 km and is 2.8 km deep[58]. This is described (yet again) as an "impact crater." Of course, of course! What else could it be! One can easily see its character is much different than other craters which are almost all circular or nearly circular. Comparing images from the websites for the USGS[59], Clementine (covered later in this chapter) and LRO[60] is an interesting exercise.

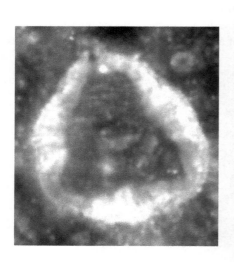

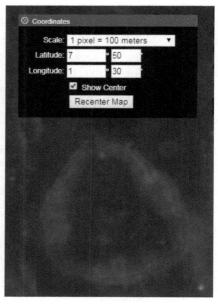

USGS image[59]   Clementine image

LRO Image[60] (on the Website, can be zoomed in further).

The Clementine imagery is next to useless – and arguably worse than an image obtained with a ground-based telescope, operated by George Tarsoudis of the Democritus Observatory[61] on 03 January 2012:

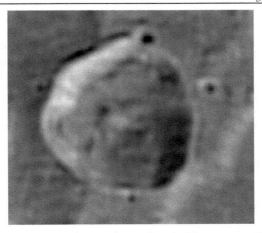

Ukert Crater image from Greek Observatory[61]

The Ukert crater has a triangular character to it, which is very unusual.

Yes, a Hexagonal Crater!

One of the most interesting Lunar images I have seen comes from Lunar Orbiter 3 and it appears to show a hexagonal crater!

Using the image data on the listing for the lo3-143-h3a image[62], I calculated this crater to be about 300 metres across. Using the co-ordinates given for this image (PP Latitude 3.3, PP Longitude -22.5) I looked for this on the LRO interactive map, but I could not find it. The Clementine image map also appears to be useless at these co-ordinates.

Lunar Orbiter Image lo3-143-h3a[63]

Chandrayaan Anomaly

Chandrayaan 1 was a lunar orbiter mission designed and launched by the Indian Space Agency (called the ISRO). According to The Times of India[64]:

> *The Rs 386-crore Indian Moon mission, Chandrayaan-1, which completes a flawless 100 days around January 30 has transmitted more than 40,000 images of different types since its launch on October 22, 2008, which many in ISRO believe is quite a record compared to the lunar flights of other nations.*

It picked up what I consider to be an image anomaly – but I cannot remember how I found out about it! A seemingly inaccurate blog posting describes it thus[65]:

> *…we looked at the one of the photo, taken by the Chandrayaan-1 Moon Mission, which comes straight from the Indian Space Research Organization (ISRO). Here's the link: ISRO official website and for our surprise one thing, the symmetry is quite remarkable, pay attention to all the mirrored details…not to mention the triangular shape: it's quite distinct. And if we believe the scale of the photograph according to its caption, 395 km., then the symmetrical object is at least 4km x 4km… a really big crater symmetry! And second one is Triangular Pyramid Anomaly. These photos (allegedly) come from the ISRO (Indian Space Research Organisation), not from NASA or the ESA.*

Unreferenced image from UFO Blogger – showing different crater field to the one below.

I say "inaccurate" because it talks about a possible pyramidal feature in the first image, but this is not referenced or linked. The Unreferenced image does not seem to show the same crater field as the lower one.

The lower image (shown below) is referenced, and the source image can be found in the internet archive. The image from the ISRO is quite interesting, but again inconclusive.

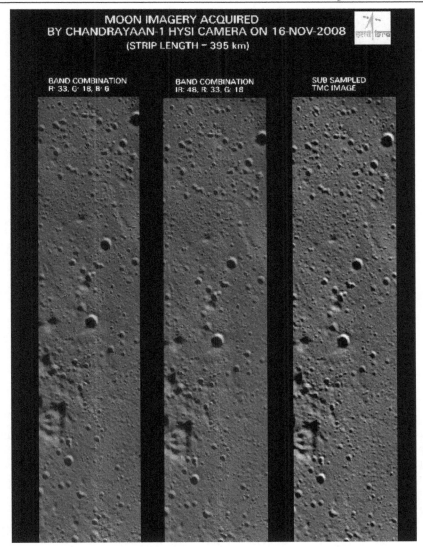

I have enlarged the interesting part below…

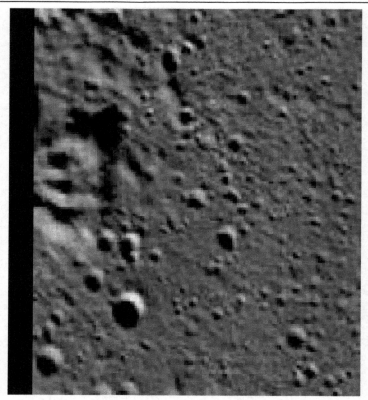

Chandrayaan Image Anomaly – 75-82 North, 5-8 West[66]

It appears that the original ISRO website has been taken offline and replaced with a new site. This site seems rather difficult to use and it is very difficult to find images - requiring use of a login (that part is easy enough) but then a Java application is downloaded, and Java security settings needed to be adjusted (it's not clear how) to get this to run. I also tried doing a basic image search in the HySI camera data. Search dates and latitude/longitude had to be manually entered and this yielded no results... Looking in the "Mosaic" section seemed to show image strips from 2009, not 2008 – and these, when downloaded, were in a proprietary format, so special software was needed to view the images...

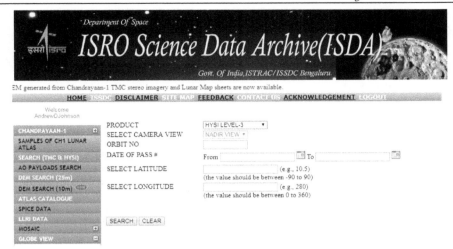

ISRO Search Page[67] – Unable to find the 2008 image!

I looked for LRO and Clementine imagery of this region, but due to the location being high in the northern latitudes, the images are taken at such an oblique angle, they are next to useless. (This is presumably something to do with the way the probe orbits the Moon). Though there are some separate polar maps from LRO, when I tried to "switch these on" to add them as layers, it didn't seem to add any new imagery at this location – perhaps because LRO has never flown over it in the way Chandrayaan 1 has.

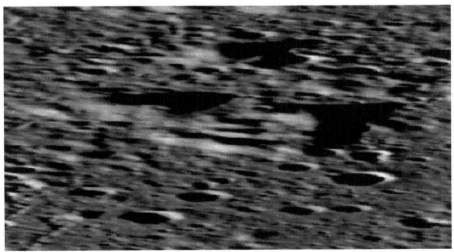

LRO image of Chandrayaan 1 Anomaly – barely discernible.[68]

The Shard and Glass Domes

In chapter four of the Hoagland/Bara book "Dark Mission", many pages are devoted to Hoagland's analysis of a number of images from Lunar missions which he claims show evidence of an enormous tower on the moon and the remnants of a gigantic glass dome. One of the objects he dubs "the shard."

In chapter 5, Hoagland claims that Apollo astronauts saw this dome when they were on the surface of the moon and that they were "persuaded" not to talk about what they had seen:

> *Early in his low-key lunar ruins investigation, a NASA insider (an MD directly involved in the medical aspects of the Program ...) confirmed to Hoagland and Johnston anecdotally that during their "debriefings" by NASA, all of the Apollo astronauts had been hypnotized—ostensibly to help them remember more clearly their time on the Moon. In actual fact, it seems more likely that these sessions were used to make them forget what they had seen, as evidenced by the astronauts own post-Apollo behavior. Neil Armstrong, for one, basically fell off the edge of the Earth, all but becoming a hermit. Besides the case of Alan Bean, which we covered earlier in this Chapter, there are at least two other cases of astronauts struggling to remember their experiences.*

The photographs that Richard Hoagland bases his theories on can be found on his website[69]. Many of these photos are Apollo mission photographs, which cannot be real, so the analysis based on them cannot be correct. Hoagland claims at various points in his book that "problems" with the Apollo image record suggest or prove that images have been altered to hide evidence of artefacts on the moon. However, based on what is discussed in chapter 17 of this book, it seems very much more likely to me that the "problems" in the images are due to them being shot in a studio.

In Hoagland's "Dark Mission" book, he returns to his "glass dome" theory several times – allegedly making his theory "provable," due to him obtaining higher quality scans of the original Apollo images from Ken Johnston and others. I agree with Jay Weidner, however, that Hoagland's dome and the supporting structure are far more likely to be parts of the studio backdrop.

The Clementine Mission

This mission is not so well known – perhaps because it was actually a military mission. The Clementine satellite measured 1.88 m in diameter and was 1.14 m long.[70] A document called "Satellite Technology" dated 1994[70] and held on the Lawrence Livermore National Laboratories website states, on page 1:

> *The Clementine satellite tested 23 advanced technologies during its mission for the Ballistic Missile Defense Organization. In fulfilling its scientific goals, Clementine provided a wealth of information relevant to the mineralogy of the lunar surface. Using six on-board cameras designed and built at the Laboratory, Clementine mapped the entire surface of the Moon at resolutions never before attained. Clementine also provided range data that will be used to construct a relief map of the lunar surface.*

I certainly find it interesting that the US Military would decide to send a probe to the moon… with all the "stresses" on defence spending, why would they spend so much money sending something over 200,000 miles to a place where there is basically nothing but barren mountains and rocks?

The LLNL document continues on page 2:

> *Some Facts About Clementine* - *Clementine is a Department of Defense program to demonstrate a new generation of technology for both military and civilian space applications. Clementine is also known as the Deep Space Program Science Experiment. It is the first in a series of technology demonstrations sponsored by the Ballistic Missile Defense Organization—formerly called the Strategic Defense Initiative Organization.*

You can browse the image catalogue on the US Navy's website[71]!

On John Lear's website, you will read claims about image "doctoring" to hide some kind of structure which was photographed in the Zeeman Crater[72]. I studied this data for a while, but found it to be inconclusive. A one-hour long video made in 2013 by YouTube user Sander19677[73] is quite well-structured and presents some interesting images and analysis, but I would say that the investigator is jumping to conclusions based on limited and "noisy" data. (Whether noise has been introduced into that data deliberately would be difficult to prove.)

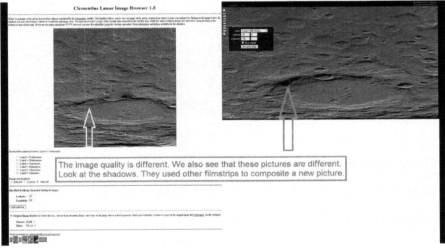

Snapshot from video, arguing that Clementine Image data has been altered[74]
You can view the current image of the Zeeman Crater by going to the "mil" Clementine image viewer website, switching on the "co-ordinates" on the Tool Panel and entering Latitude -71 and Longitude -138:

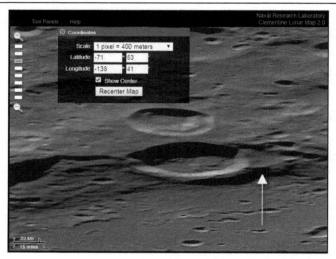

Alleged doctoring in Zeeman Crater imagery from US Military Clementine mission[71]. The area which is roughly triangular in shape is alleged to contain an enormous structure – about 3½ miles in length!

Colour images of the Moon?

A common theme that we shall see is that the majority of close range images of the surface of the planets and the moon that have been presented to the public are in black and white. However, there are some colour images of the moon in the LRO archive[75]. An example image I found below shows a very bright, almost perfectly circular feature which has a small blue colouration in one area:

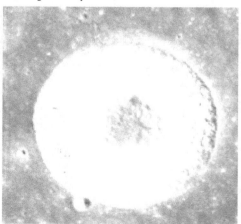 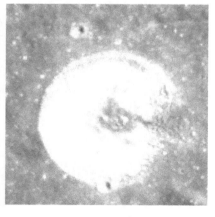

LRO image showing blue colouration in the "8 O'clock" and "9 O'clock" positions[76]. This feature is about 16 km in diameter.

Another nearby feature - about 12 km in diameter.

Conclusions

When I started writing this chapter, I had the impression that the evidence for lunar anomalies I had seen did not show anything that really jumped out and "punched me in the brain." Having finished writing this chapter, I have spent a considerable number of additional hours trying to find out more information about the anomalies, and more imagery of them. Though I have found some more images and a little analysis of them, I can't say that I feel I have seen anything which looks to have been "obviously constructed."

Having spent quite a few hours poring over LRO images, it is worth stating something that is probably fairly obvious. The way that objects are illuminated can considerably affect the pattern which appears – and it is exactly this argument which has been used to "explain away" more compelling evidence. That is to say, experts will just describe it as "an effect of light and shadow." I think this is probably true of a fair few images in George Leonard's book – as some of the artefacts can be identified when better resolution imagery is available or with the same scene being illuminated differently.

From the lunar images I have shown here, I think the strangest anomaly is the hexagonal crater – and I don't seem to be able to find a lot of information about that. Neither does it appear to be in the LRO imagery, unless the original location given for the object was incorrect.

Websites which display and discuss these images generally fall into one of two categories. (A) They will claim (or strongly imply) that they are evidence of some type of alien or intelligent activity (as is the case in books like George Leonard's) or (B) they will contain comments which, in some way[77], mock the images or those people who study them. Much less common is an attempt to explain what these images actually show and how the structures in them might have formed. Large scale hexagons, as far as I am aware, are rare in nature. Small hexagons are very common, however (as seen in a honeycomb, basalt columns, crystals and other structures).

In the next chapter, we will "fly to" Mars and look at what I consider to be much more compelling evidence of things that have been constructed.

4. A "Face-off" on Mars

There are a number of Web Sites which have described numerous Mars anomalies, such as the now defunct website of Dr Tom Van Flandern http://www.metaresearch.org and Joe Skipper's site http://www.marsanomalyresearch.com/. However, some images on these sites are difficult to "get excited about." In addition, hundreds of YouTube videos have been produced by people interested in finding things in NASA images, which have seemingly been overlooked. We could fill several volumes with a discussion of some of these videos - I have watched a few of them and, often, I simply cannot see what is being claimed – or I think the observer has mistaken the object for some type of image artefact or other effect.) If you can find anything that is <u>fully referenced</u> and you feel it is more compelling than what I have covered in this volume, feel free to contact me!

Mars Facts

Mars has been a subject of intense interest in unmanned space exploration. To date, there have been over 50 missions which have been aimed there, a fair number of which have failed or been aborted. Science fiction stories and films/movies about the Red Planet have been published at regular intervals and, since at least the 1970s, there has been ongoing discussion and plans for manned missions there.

As can be seen from the table below, Mars is roughly half the size of the earth, has one third of the gravity at the surface and is somewhat less dense.

	Mars	Earth	Ratio (Mars/Earth)
Mass (1024 kg)	0.6419	5.9736	0.107
Volume (1010 km³)	16.318	108.32	0.151
Equatorial radius (km)	3397	6378	0.533
Polar radius (km)	3375	6356	0.531
Mean density (kg/m³)	3933	5520	0.713
Surface gravity (m/s²)	3.69	9.78	0.377

Using Google Earth (Mars) To Look for Anomalies

If you are a curious person and have a computer which has a 2 GHz processor or faster and 4GB of RAM or more, you should be able to download Google Earth Pro and then switch to a view of Mars. You can then enter the latitude and longitude figures to find areas of interest and even "fly" over them and study them in more detail.

Dr Carlotto produced a short guide to help you to do this[78], although be aware that it may be out of date, as software is updated, features are added, moved and removed. If you are going to use Google Mars, make sure you have the "CTX Mosaic" selected in the "layers" selection pane.

The Viking Missions

The story of Mars anomaly research became much more public during the Viking Missions[79]. Viking 1 and Viking 2, each consisted of an orbiter and a lander. The primary mission objectives were to obtain high resolution images of the Martian surface, characterize the structure and composition of the atmosphere and surface, and search for evidence of life. Viking 1 was launched on August 20, 1975 and arrived at Mars on June 19, 1976. Viking 2 was launched on Sep 9, 1975 and arrived at Mars on 08 July 1976. The total cost of the Viking project was roughly one billion dollars.

A Face, "Staring into Space…"

So, let us now discuss some of the story of what was apparently one of the most amazing discoveries of the first 15 years of Mars exploration. If there is one anomaly in the Solar System that people seem to have heard of, it is the so-called "face" on the surface of Mars, discovered by the Viking 1 orbiter in 1976. I will not be discussing this anomaly in great detail, because numerous books have already been written about it, so as I mentioned earlier, I suggest the reader gets hold of one of those. A useful book is the one by Mark Carlotto called "The Cydonia Controversy[23]". (Cydonia is the name given to the area of Mars where the face is located – and the reason for that name being given is itself a topic of interest. Cydonia was a "city-state" on the island of Crete[80], but the region of Mars had been named many years before the Viking missions. Cydonia had been named because of its appearance in Earth-based telescopes.)

It appears that there is something important about the Face which NASA, ESA and other agencies don't want us to know. I say this because it is clear that a

NASA scientist was willing to lie about facts relating to the Face's discovery. I will try to explain this shortly. But here is what NASA reported to the world on 31 July 1976[81]:

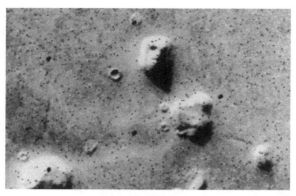

Geologic 'Face on Mars' Formation - Viking 1 Orbiter Face on Mars July 31, 1976 (Image No 035A72)

NASA's Viking 1 Orbiter spacecraft photographed this region in the northern latitudes of Mars on July 25, 1976 while searching for a landing site for the Viking 2 Lander. The speckled appearance of the image is due to missing data, called bit errors, caused by problems in transmission of the photographic data from Mars to Earth. Bit errors comprise part of one of the 'eyes' and 'nostrils' on the eroded rock that resembles a human face near the center of the image. Shadows in the rock formation give the illusion of a nose and mouth. Planetary geologists attribute the origin of the formation to purely natural processes. The feature is 1.5 kilometres (one mile) across, with the sun angle at approximately 20 degrees. The picture was taken from a range of 1,873 kilometres (1,162 miles).

So that settled it – the "Face" was simply a combination of light and shadow, in a low-resolution image. The appearance of a face in the image was because of our pre-disposition for pattern recognition. We have all seen "faces in the clouds" and we have "the man in the moon" etc – it was just another version of that… What else would a "proper scientist" have to say about it?

If you read Dr Mark Carlotto's book, you will see that Dr Jerry Soffen lied about the data. I have to emphasise that this is my characterisation of Soffen's statement, not Dr Mark Carlotto's. Soffen claimed it "was just a trick of light and shadow" and when the Viking Probe returned several hours later and took another photograph, the effect had disappeared (because of different lighting conditions etc). He lied because "several hours later," the Viking Probe was thousands of miles away from the face and so could not have taken a photo! This was not confirmed until 2 or 3 years later when someone went to look for the "second image" of the face. An honest person would have said something like *"sorry, I was wrong – we didn't take a second image hours later. It came much later. At the moment, I don't know what the 'Face' is, but presume it is a natural feature."* Soffen never said such a thing – and we will learn, in a short while, what he did say, later on, about the Face.

Vincent Dipietro, then a Systems Engineer at the NASA Goddard Space Flight Centre[82], first saw the image of the Face in a magazine article in 1977 and took it to be a joke. Two years later he came across the same image - in the photographic archives of the National Space Science Data Center (NSSDC) at the Goddard Space Flight Center outside Washington DC (where he was working).

> *"There before me in black and white was the same serene image of a human-like face against the background of the Martian land surface. The title was certainly not misleading; it simply said 'HEAD'...*
>
> *"At this point I knew the object was not a hoax or it would not have been so boldly displayed in the NASA archives. I felt relieved and inquisitive; relieved that NASA had noted the picture and would presumably have verified it, and inquisitive to want to know more. But there was nothing more."*

This was eventually found in the archives:

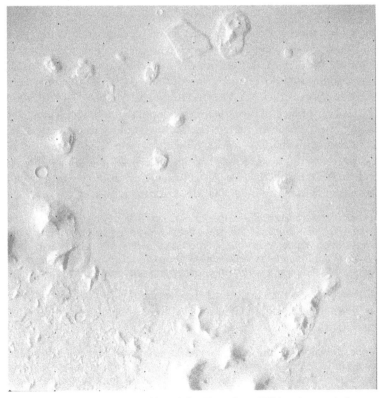

The second image – no. 070A13[83] - of the Face from Viking (upper left corner) – showing different lighting and orientation respective to the Viking. This image was also taken in 1976, but the date is unknown[84].

Let's put these 2 images side by side, without any processing other than cropping and rotating.

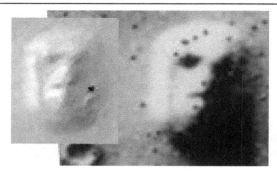

Crop from Image 70A13 on the left and Crop from 035A72 on the right

It is easy to see why this became an image of intense interest to some people.

Since 1976, the Face has been photographed several times – most recently by Mars Reconnaissance Orbiter, which we will cover later. For the moment, we will remember that we are dealing with <u>low resolution</u> images of the face.

Just pausing for a moment, we can observe that we almost certainly have many more data points on the Face image than we do on the Blair Cuspids image from the moon. Indeed, Mark Carlotto did a similarly detailed analysis on the face images and he used different techniques. He describes this in a section of Chapter 5 of his book called "A Trick of Light and Shadow?" He describes how he and others at his place of work were developing a technique called "shape-from-shading" (or photoclinometry). This allowed them to generate a 3D image from a 2D image with shading. Carlotto includes more information on his own website (carlotto.us[85]). He describes how their technique was tested on known objects before being applied to images of interest. Before showing a selection of 3D renderings generated by their method, Carlotto writes:

> *The results were provocative: the computed 3-D structure clearly showed evidence of facial features in the mesa itself. Moreover, by generating views for other light source positions (other times of the day and year), and for other viewpoints (from the side as well as above), we could show that the appearance of a face persisted over a wide range of lighting and viewing positions.*

Carlotto describes a second technique they applied to the images, which related to Fractal Geometry. Again, in Chapter 5 of his book, in the section entitled "Measuring Artificiality," he writes:

> *Generally speaking, nature tends to create structures that are 'self-similar' — structures in which a part resembles the whole in some sense. For example, over a range of scale, as one examines a leaf in greater and greater detail, the same kinds of patterns repeat at smaller and smaller scales. Objects that exhibit this kind of behavior are known as fractals. Fractal models can be used to describe a variety of natural phenomena besides plants, including clouds, lightning, and the shape of natural terrain and coastlines, to name just a few...*

He continues:

> *A colleague of mine, Michael Stein, had developed a fractal technique for detecting manmade objects such as military vehicles in overhead imagery (Figure 18). Without any modification we applied the technique to Viking frame 35A72, and found that the Face was the least fractal object in the entire image!*
>
> *Extending our analysis to other nearby frames, the Face remained the least fractal object over an area about 15,000 sq. km in size — more than five times the area of the state of Rhode Island. Results from Viking frame 70A13 corroborated this result.*

Hence, their algorithms suggested that the face was, indeed, artificial.

Carlotto also describes other features in the Cydonia Region, such as the D & M (Dipietro and Molinar) Pyramid, which we will look at later. Carlotto also goes into some detail about how a team of independent scientists came together to study the Viking Images. The group was called The Independent Mars Investigation team. The team was formed largely as a result of the efforts of former NASA curator, Richard C Hoagland. Members formed a loose association and wrote papers, some of which were eventually published in the Journal of the British Interplanetary Society. Carlotto's results were eventually published there in 1990 in a paper called "A Method of Searching for Artificial Objects on Planetary Surfaces." Dr Brian O'Leary – former Mars Mission astronaut - had also worked with the group and had his own papers published, but had at least one rejection by the journal.

What Did Dr Jerry Soffen Know?

In Richard Hoagland's "Dark Mission" book, Hoagland, reports that a designated Viking Lander site was changed from Cydonia after orbiter pictures were obtained. This is confirmed in an article on one of NASA's websites[86]:

> *Results of the landing site search were made public on 7 May 1973. John Naugle announced that a valley near the mouth of the six-kilometer- deep "Martian Grand Canyon" was the target for the first lander. Known as Chryse, the region had been named for the classical land of gold or saffron of which the Greeks had written. If all went well, Naugle told the assembled press corps, the first Viking would be set down on or about the Fourth of July 1976. The backup to the Chryse site was Tritonis Lacus, Lake of Triton named for the legendary river in Tunisia visited by Jason and the Argonauts. The second Viking was targeted for Cydonia, named for a town in Crete, with Alba, the White Region, as backup. **Soffen told the press that NASA hoped Viking 1 would be heading for a very safe but interesting target.** The scientists had decided early that the first site should be sought in the northern hemisphere (because it would be Martian summer there), at the lowest elevation possible (higher atmospheric pressure and better chance of water in some form), on the flattest, least obstructed region they could find (for landing safety and weather observations). But the second mission had been a different story. The biologists wanted water, and after much debate and study they hoped to find it in the 40° to 55° north latitudes. Their Mission B sites were just above 44°.*

Is this the reason Soffen lied about finding a second image of the face? Could it even be that the "second" image of the face (070A13) was actually taken first, and it was after seeing this image that the landing site was changed?

NASA Shows no Interest

Despite the work of O'Leary, Carlotto and others, indicating the Face might be an artificial structure, NASA remained uninterested. Viking mission scientist Dr Jerry Soffen, who originally announced the discovery of the "Face" was asked at the July 1985 "Steps to Mars Conference" in Washington DC, by journalist Jeff Greenwald about the independent Mars research that had been going on. He replied

I really haven't been that interested; and I'm still not.

He had an opportunity to correct his earlier statement about a "second photo" of the Face – which couldn't have been taken, as he'd said years earlier. However, he chose to let the lie he told stand.

At the same meeting, Carl Sagan said to Jeff Greenwald[87]:

What they [the Independent Mars Investigation Team] are proposing is to use existing – and some novel – computer enhancement techniques on existing data. Now, this area, like all areas of Mars, has already been subject to state-of-the-art enhancement... And there's nothing that comes out beyond what you've already seen... I'm not opposed to investigating. My view on the Face on Mars is my view on astrology. If someone can show that there is some validity to the claims, that's useful. But since the vast preponderance of the evidence is that it's nonsense, I don't think that's a good investment of resources.

Mars Orbiter - 1992

In 1990, plans for a new Mars orbiter were well underway. Mars Orbiter was launched on 25 September 1992[88]. A new high-resolution camera had been designed by Michael Malin and he was contracted by JPL to "design, develop and operate instruments flown on manned and unmanned spacecraft." For this reason, he formed Malin Space Science Systems – this made things a lot cheaper for NASA, as the camera operation and image acquisition and processing could be streamlined. However, it meant that Michael Malin would be in control of all the images that came back from the new orbiter. Mark Carlotto writes in "The Cydonia Controversy"[23]:

Malin was not required to release the imagery to the public until after a six to nine month 'data validation' period.

Around this time, the Mars Investigation Team tried to apply pressure to NASA to make sure they would take more images of the Cydonia region during future Mars missions. With this in mind, they produced a report.

The McDaniel Investigation

In autumn 1992, Stanley V. McDaniel, a philosophy professor at Sonoma State University, California, became interested in the Cydonia investigation and he had reviewed some of the independent literature. He soon found out NASA had not really published anything of note about the issue and he generally wasn't happy with the answers NASA was giving in relation to the issue.

McDaniel Report – making recommendations about NASA's Mars Research Programme.[89]

It was published in 1993 - seventeen years after the NASA's Viking Orbiter first imaged the Face, as the new Mars Observer probe was approaching the Red Planet. The report was used to apply political pressure to NASA, via congress, to set mission priorities to include re-imaging Cydonia. Both Richard C Hoagland and Dr Mark Carlotto go into more detail about the McDaniel report. McDaniel was scathing in his description of NASA's reaction to the independent Mars research. He wrote:

> *However, during the seventeen years since the controversial landforms were discovered, NASA has maintained steadfastly that there is 'no credible evidence' that any of the landforms may be artificial. A close look at NASA's arguments reveal that NASA's 'evaluation' has consisted largely of initial impressions from unenhanced photographs, heavily weighted by faulty reasoning…*
>
> *NASA has failed to apply any special method of analysis; it has relied upon flawed reports; it has failed to attempt verification of the enhancements and measurements made by others; and it has focused exclusively on inappropriate methodology which ignores the importance of context…*
>
> *There remains no scientific basis for NASA's position regarding the landforms…*

In the report, McDaniel expressed dismay and openly criticised Carl Sagan:

> *… my original naive view – that all NASA scientists were sincerely interested in the truth – was utterly shattered when I discovered the most blatant piece of disinformation I have ever seen: one written not by an obscure NASA Public Information employee, but by a prestigious member of the 1976 Viking Lander Imaging Team, Dr Carl Sagan. Dr Sagan's contribution to the subject could not*

be interpreted as mere scientific bungling; its author is too knowledgeable for that.

He praised the work of the Independent Researchers.

As my study of the work done by the independent investigators and NASA's response to their research continued, I became aware not only of the relatively high quality of the independent research, but also of glaring mistakes in the arguments used by NASA to reject this research...

With each new NASA document I encountered, I became more and more appalled by the impossibly bad quality of the reasoning used. It grew more and more difficult to believe that educated scientists could engage in such faulty reasoning unless they were following some sort of hidden agenda aimed at suppressing the true nature of the data.

McDaniel went so far as to suggest:

The concept of withholding information on a possible extra-terrestrial discovery conflicts with an understood NASA policy to the effect that information on a verified discovery of extra-terrestrial intelligence should be shared promptly with all humanity.

Additionally, he referred to

...the possible sequestering of the data under the aegis of "private contract" ...

and

...a sense of complacency around the issue all support the suspicion of a motivation contrary to the stated policy...

Further extracts from the report can be read online[90]. It would be several more years before the pressure to re-image Cydonia resulted in new images being acquired. Contact with Mars Observer was lost on 21 Aug 1993 and was never re-established.

In chapter three of their book "Dark Mission," Hoagland and Bara discuss a few reasons why they think the Mars Observer mission "went black." They suggest that, due to what the independent Mars Investigation Team was exposing about Cydonia and the public interest they had aroused, the Mars Observer mission was "taken over" by the military for their own purposes – and the "communications loss" was a cover story. Hoagland/Bara point out that this happened at exactly the same time as Hoagland was involved in a televised debate regarding Cydonia anomalies with Mars Observer mission director Bevan French, on 22 April 1992. (Hoagland claims he "won" this debate hands down.) Though there is no proof that the two events are connected, it would not surprise me at all if the military do, indeed, know a lot more about Mars and Cydonia than we do.

Mars Global Surveyor Photographs the Face in 1998

After the loss of Mars Observer, plans to build a new probe, using most of the technology from it, were drawn up. Mark Carlotto discusses this in some detail in "The Cydonia Controversy"[23]. On 7th November 1996, Mars Global Surveyor was launched and it achieved orbit on 12 September 1997, though adjustments to the orbit continued to be made for over a year before the surface of Mars was systematically photographed.

As reported in "The Cydonia Controversy"[23], following publication of the McDaniel report, McDaniel formed the Society for Planetary SETI Research (SPSR), a multi-disciplinary group of about two dozen scientists and other professionals interested in the search for, analysis and evaluation of possible evidence of extraterrestrial artefacts in the Solar System – with Cydonia being the main area of interest at the time. On 24th November 1997, six representatives from SPSR: Mark Carlotto, John Brandenburg, Horace Crater, Vincent DiPietro, Stanley McDaniel and David Webb, met with Carl Pilcher the Acting Director of Solar System Studies, and NASA scientist Joseph Boyce[91]. Following this short meeting, David Webb of the SPSR reported that NASA had acknowledged their requests, at least to some degree. Webb said that Carl Pilcher (the Director of Solar System Studies):

> *... firmly and unequivocally stated that it was the official policy of his department and of NASA that the Cydonia region was to be imaged at high resolution during fly-over and that Glenn Cunningham and Mike Malin were aware of, and had signed off on the policy... He went on to say that everyone was interested in having the area imaged: one group because they wanted to show us how wrong we are and have been all along; the other group, because they feel that we have some interesting material, and they would like to see just how interesting it turns out to be.*

In chapter 8 of Mark Carlotto's book, he describes how he eagerly awaited the new images from MGS and how bad they appeared to be. When I first saw these, they seemed, to me, to be deliberately bad. They seemed to be heavily processed to make the landscape look dull and almost featureless. The releases can be found on the Malin Space Science Systems website:

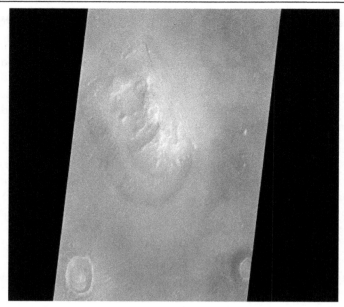

Calibrated, Mercator map-projected (flipped left to right), contrast-enhanced, filtered image.[92]

Mark Carlotto writes how he spent considerable time studying this image and realised that the Face had probably been photographed through thin cloud/haze, which had made shadows weaker. The angle from which the photograph had been taken by the orbiter was also quite low. After weeks to months of studying the new image, Carlotto concluded that features of a face were still present.

In his book, Carlotto discusses the reaction to his research, which was a mixture of ridicule and anger. Carlotto had been contacted by the Journal of Scientific Exploration for an opinion on the new image. Carlotto responded with a draft paper he had been developing. The JSE editor declined to publish this and, instead, published an opinion piece by Viking Mission Scientist David Pieri, which contained this statement[93]:

> *the ludicrous allegations of conspiracy and data suppression that followed over the years were particularly galling and scurrilous. Those of us on the Viking orbiter and Lander Science Teams would have liked nothing more than to trumpet to the world the discovery within our data of the evidence of an extraterrestrial civilization. But we just didn't see it. I firmly believe that someday—in this Solar System or in some other—man will inevitably encounter such evidence, but, for me, the Face on Mars isn't it.*

The piece made no mention and had no critique of the earlier research by Carlotto and others regarding such things as the 2D and 3D features of the face, identified in the "shape from shading" study. Readers might want to compare figures 7a and 7b from the Pieri paper to the ones created by Carlotto:

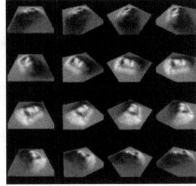

Pieri / Carlotto face modelling comparison

Non-Random Mound Configurations

In 1999, Prof Stanley McDaniel and professor of Physics at the University of Tennessee's Space Institute, Horace Crater wrote a paper called "Mound Configurations on the Martian Cydonia Plain[94]." It was published in the Journal of Scientific Exploration in 1999 [Vol 3, No 3, 1999]. In that paper he wrote:

> *"Investigation of the geometric relationships between these mounds takes the form of a test of what may be called the random geology hypothesis. The hypothesis presupposes that the distribution of the mounds… however orderly they may seem, is consistent with the action of random geological forces…*
>
> *Our question is: Does the random geology hypothesis succeed or fail in the case of the mound configuration at Cydonia?"*

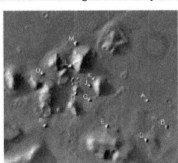
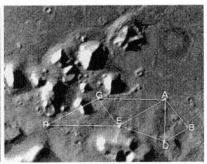

Randomness of the position of these mounds was tested by an algorithm in the McDaniel / Crater paper.

Rather than select the mounds to analyse, software was written to search for mounds, to see what alignments between them could be found. At 5% precision, nothing significant showed up, but at 0.2% precision, it was found that lines drawn between mounds formed triangles in a less random way. In the paper, they conclude:

We must conclude that the random geology hypothesis fails by a very large margin, that a radical statistical anomaly exists in the distribution of mound formations in this area of Mars. Since previous research in this area seemed to indicate possible anomalies (including, but not limited to the controversial Face), we had reason to focus on this region... The existence of this radical statistical anomaly in the distribution of mound formations in this area of Mars indicates, in our opinion, a need for continued high priority targeting of the area for active investigation and determination of the origin and nature of the mounds.

Reviewers of the paper made a number of objections to it. Crater and McDaniel responded to those objections. Here is one of the objections and their response to it:

The authors use the term "random geology hypothesis." This is an unfortunate term to use. Some geological formations are obviously very far from random. (Consider, for instance, the "Devil's Post Piles" in Ireland.) The authors are really testing whether the distribution is random. If they were to show convincingly that it is not, they may simply have shown that this array of mounds is another geological formation that is non-random. Of course, that in itself could be interesting.

They responded:

The objection seems to assume that non-random geological formations are common, but this is not supported either qualitatively or quantitatively, nor is the method by which it would be determined stated. We are not aware of any studies, comparable to our own, carried out on terrestrial formations, that would identify such a degree of non-random distribution of separate objects. In discussions with geologists we have found none who could identify distributions of natural objects in terrestrial terrain displaying such non-random characteristics. Our research poses the question "is there any known geological mechanism that could produce such coherent patterns among mound-sized objects over distances of kilometers?" Thus far no adequate geological answer has been forthcoming. Archaeologist J. F. Strange of the University of South Florida has stated on the basis of our data, and on the basis of his separately applied Kolgorov-Smirnov test, that if the mounds had been found on the Earth, archaeological teams would be strongly motivated to investigate them. We therefore believe our results are substantive and indicate that further serious study of the region should be undertaken.

Despite these detractors and debunkers, over the years Dr Mark Carlotto, Horace Crater and Prof Stanley McDaniel continued their research and efforts to encourage NASA to obtain more data.

Alignments in Cydonia

In both Richard C. Hoagland's book "Monuments of Mars" and Mark Carlotto's book "The Cydonia Controversy", some time is spent on the way that mounds/surface features/monuments are aligned. I would agree with the general observations about these alignments, which seem to indicate the features are archaeological rather than geological in nature. I would also advise caution in relation to taking everything Richard C Hoagland says as accurate. Much as I am grateful to him for helping to expose some of the information about Cydonia, Mars and other features seen elsewhere in the Solar System, I know that he has helped to put out disinformation about the events of 9/11. (For details, please see the article on my website "Is Richard Hoagland on a "Dark Mission"?[95] or my free eBook "9/11 Finding the Truth"[96]. We will further discuss Richard C Hoagland's role in chapter 23.)

The Face... and an Exploding Planet?

A further interesting consideration is made in Carlotto's book about the location of the Face. Dr Tom Van Flandern (1940 – 2009), formerly of the Naval Observatory[97], was the main proponent of the "Exploded Planet Hypothesis," (EPH)[98]. This hypothesis argues that there was a major planet in the region of the asteroid belt, in our Solar System. Van Flandern suggests that the ensuing Solar System cataclysms caused by an exploding planet could have caused the polar orientation of Mars (and perhaps Uranus) to have shifted. This might also explain the appearance of moons like Miranda[99].

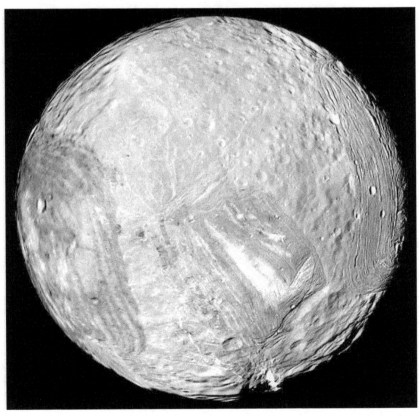

Miranda – a moon of Uranus – appears to have broken up and re-formed.[100]

Based on some of the geological evidence on Mars, it is suggested that before the polar shift, the "Face" would have been near the Martian equator[101].

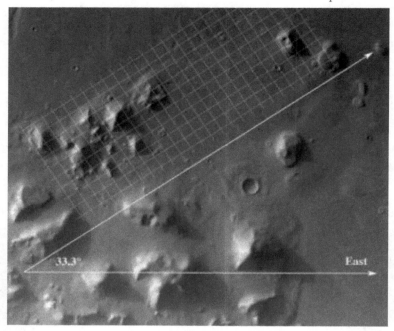

Face's position in relation to the old Mars Equator?

Death on Mars?

Another of the researchers in the Mars Investigation team was Dr John Brandenburg. He suggests that the civilisation on Mars was destroyed in an ancient nuclear war.

Dr Brandenburg, who has a PhD in Theoretical Plasma Physics from the University of California at Davis, has proposed that a Martian civilisation may have been wiped out by a thermonuclear explosion many thousands or millions of years ago. This idea isn't that new, because when Mars was first photographed in the 1960s and 1970s, it became clear it was mainly desolate, but unlike the Moon, it had an atmosphere. In March 2015, Brandenburg was featured in an article in the UK Daily Mail[102] quoting him as saying:

'We have now found evidence of the nuclear melt-glass Trinitite (formed on Earth at the site of nuclear weapon air-bursts) at both sites of the hypothesised explosions and this will be presented at the conference,' Dr Brandenburg revealed to MailOnline ahead of his talk.

'This strongly supports my hypothesis of massive nuclear airbursts.'. 'So far no scientist has offered any other explanation for this body of data.'

My own opinion on this is that Brandenburg's evidence is quite thin. Indeed, if you have a look at a presentation he gave[103], there only seem to be a couple of items that he can site as evidence. (Brandon gave a version of this presentation at the Secret Space Program Conference in Oct/Nov 2015, in Bastrop Texas[104].) These pieces of evidence are the melt-glass, mentioned above, the detection, in the Martian atmosphere of a higher proportion of an isotope of Xenon – Xenon 129 compared to what has been seen elsewhere in the Solar System. This isotope, it is said, is only produced in nuclear fission events – such as the detonation of nuclear bombs. Brandenburg argues, therefore, that any Martian civilisation may have been wiped out by a nuclear war.

Exploded Planet?

However, Tom Van Flandern states the presence of this isotope could be the result of the effects of a nearby exploding planet.

> As further evidence of this scenario, we note that Mars has an anomalous ^{129}Xe content in its atmosphere that is nearly triple that found on other bodies where it has been measured (DiPietro, 1996). Since ^{129}Xe is a second order nuclear fission by-product and does not arise through normal nucleosynthesis, it has long been assumed that an ancient supernova was responsible for the presence of that isotope in the Solar System. Then why does **Mars** have an anomalously high amount of it? Again, its proximity to the eph event is an obvious explanation for Mars in particular to be anomalous.

Stephen Smith, a proponent of the Electric Universe theory/model, argues that the Xenon 129 isotope could have been created in Mars' atmosphere during electrical discharge events - which are also said to have caused the formation of many of the surface features on Mars[105]. Brandenburg responded to this suggestion thus[106]:

> I have seen no evidence of big electric arcs in space except at the moon Io orbiting Jupiter. I have however seen video and data from big nuclear weapon tests and went to grad school at a national lab where they did fusion research at one end and designed nuclear weapons at the other.
>
> So I tend to offer explanations based on known and well characterized phenomenon. Mars has a very weak magnetic field and thus is a poor prospect for any big electro-dynamic phenomena leading to planet-rending electric arcs, in my opinion. So as outlandish as my hypothesis is, it at least invokes known phenomenon. For that reason I think it is more likely to explain what is seen on Mars.

Overall, then, even if people could agree there was an advanced civilisation on Mars at one time, I don't think there is enough evidence available to us to say how this civilisation was destroyed – or even that it *has* been destroyed. Perhaps there are underground bases, who knows... Now, all we can see on the surface seems to be evidence of desolation. Maybe someone does know what happened to this civilisation. We will return to this theme in a later chapter.

MGS Images the Face Again - in April 2001

MGS continued to return high resolution images, a few more of which we will discuss later. Mark Carlotto also discusses them in "The Cydonia Controversy"[23]. Further images of the "Face" were posted on the MSSS website on 24th May 2001[107]. Coincidentally, this happened after a man called Peter Gersten wrote to NASA asking them to release all images of Cydonia and that further images be acquired as a matter of priority, when the opportunity arose. On 11 May 2001, Edward J. Weiler, Associate Administrator for Space Science at NASA replied to Gersten stating that Cydonia/Face images that had been taken on April 8, 2001 had been released "via multiple public websites."[108] Unfortunately, no one could find them until almost two weeks later – on the 24th May 2001! NASA posted a page called "Unmasking the Face on Mars"[109]. It states:

> What the picture actually shows is the Martian equivalent of a butte or mesa -- landforms common around the American West. "It reminds me most of Middle Butte in the Snake River Plain of Idaho," says Garvin. "That's a lava dome that takes the form of an isolated mesa about the same height as the Face on Mars."
>
> Cydonia is littered with mesas like the Face, but the others don't look like human heads and they've attracted little popular attention. Garvin and other members of the MGS Science Team have studied them carefully, however, using a laser altimeter called "MOLA" on board Mars Global Surveyor.
>
> MOLA can measure the heights of things with a vertical precision of 20 to 30 cm (its **horizontal resolution is 150m**). "We took hundreds of altitude measurements of the mesa-like features around Cydonia," says Garvin, "including the Face. The height of the Face, its volume and aspect ratio -- all of its dimensions, in fact -- are similar to the other mesas. **It's not exotic in any way.**"

Mark Carlotto analysed the May 2001 images in his book, so I direct the reader there to find out about his interesting and expert conclusions. He also notes that the MOLA (height) data is too low in resolution to be useful on the face. The face is approximately 1km in length and so would be less than "10 elements" long in MOLA data.

Although Carlotto revised some of his conclusions about some features of the face – now that higher resolution images were at last available, his overall conclusions remained the same. It was "not obviously" a natural formation. He considers the effects of erosion on the features, the most obvious part of which, to me, seems to be in the Face's lower left (to the right, in the image below) and its "pedestal." Carlotto rightly points out that if we were to look at some archaeological features on earth – such as some of the older, very eroded pyramids, it can be very difficult to distinguish them from *geology*. This distinction of course, becomes much more difficult when close-up, personal inspection is not a realistic option.

For myself, I would strongly argue the 2001 image is a good match to the morphology of a human face. However, if we disregard our tendency to anthropomorphise, just consider the base, I am in complete agreement with Carlotto that we can observe a high degree of symmetry.

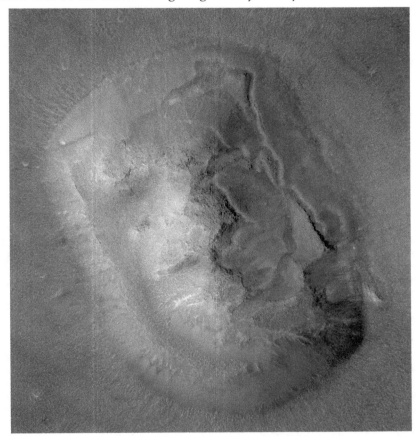

MGS MOC Release No. MOC2-283, 24 May 2001[110]

Cydonia – 2001 Press Conference

It was only recently that I discovered that, on May 8, 2001 at the New Yorker Hotel, a press conference was held regarding the MGS images of the "Face." [111]This was only about 2 weeks before MSSS posted the clearest images of the "Face". This press conference was also, coincidentally, the day before Dr Steven Greer's Disclosure Project Press Conference, where over 20 highly qualified military and civilian witnesses stated they would testify before congress in relation to their UFO/ET-related experiences[112].

The conference in New York was introduced by Michael Luckman – author of the book "Alien Rock" and a long-time activist in the field of UFO/ET research[113]. Luckman introduced two scientists we have already mentioned – Dr

Tom Van Flandern and Dr Brian O'Leary (27 Jan 1940 – 28 July 2011). Michael Luckman described the release of new NASA images of the "Face" as a part of a wider disclosure, but Van Flandern and O'Leary did not seem to see it this way.

From Left to Right: Michael Luckman, Dr Tom Van Flandern, Dr Brian O' Leary

In his short speech, O'Leary stated:

More than two decades of responsible research by investigators outside of NASA on this object - the so-called Face on Mars, have revealed very strong evidence - not yet proof - that this object and other objects nearby were artificially constructed and yet NASA and its contractors have shown no interest in these objects and have obfuscated the research.

In the Question and Answer session, O'Leary stated:

I was deputy team leader of the Mariner 10 Venus/Mercury mission. Mike Malin was in charge of the imaging for the Mars Global Surveyor [and he] was a graduate student then. He basically worked for me and since I began to give credibility to the inquiry - not necessarily the result - then all of a sudden, I was out. Not one penny from NASA, not one acknowledgement! This is what's happening culturally. And if you understand that, then you begin to see the pattern [that] the geologists on the Mars Global Surveyor team look at Mars through geologically colored glasses. So they're going to be looking for canyons and signs for water erosion and so forth and so on. Where some of us outside of NASA have "culturally coalesced" - admittedly in a kind of an ad hoc way - because we haven't been organized or funded for any of the work. And so if you look at it from that point, you begin to understand how it might be that these investigations actually are every bit as credible as the ones that happen within NASA.

Van Flandern discussed several anomalies, including the Face and a "T" shape, which we will see in chapter 6. Of the Face and the Cydonia region in general, Van Flandern stated:

All the features showed up and they have the right size, shape, location and orientation. [This] gives us very large combined odds against chance and the fact that there is no background of similar features means that the calculations are statistically significant, to a scientist. The combined odds of this feature arising as a product of nature or "the chance origin hypothesis" are a thousand billion billion to one. The artificial reality of Cydonia is therefore established beyond a reasonable doubt.

Other features he showed, however, seemed much less compelling and don't seem to have as many associated measurements (suggesting specific geometry) as the Cydonia region does. Neither did they seem to have been subject to fractal analysis or "shape from shading." Were these "thrown in" to dilute the presentation?

Around the time of this press conference, there was more of a disagreement between Hoagland and Van Flandern than in years previous.

Hoagland began to argue that the "Face" had feline features and in the "Dark Mission" book, he goes to some lengths to suggest this feline nature is quite obvious (I disagree). I think Hoagland makes too much of the link to the Sphinx in the Face's configuration. Nevertheless, I do agree with him that the significance of Egyptian symbolism should not be ignored. This symbolism seems to be more pervasive than most people realise. (Interested readers can watch Scott Onstott's amazing video series "Secrets in Plain Sight" [114]to learn more.)

Getting Messianic?

Quite a few interpretations of the meaning and true nature of the artefacts in Cydonia can be found online. Some of them go way beyond what seems acceptable in a logical analysis. One such interpretation, by "Max the Knife" called "The Crux of Cydonia"[115] includes some very detailed graphics and quite a long narrative. The contents of this 97-page document (all presented in a somewhat difficult to read "landscape" format) are, to me, marred in the final sections headed "Is it possible that the Messianic Prophecies are all about me?!" The author then presents facts about himself which seem to "line up" with certain facts about Cydonia. The author references some of Hoagland's work and that of George Haas, who together with William Saunders authored a poorly-regarded book called "The Cydonia Codex."[116] The book argues that the Face is at least partly feline. (I heard about this book at the time it came out, but I wasn't moved to study it then, or now.) As mentioned earlier, Richard C Hoagland also (unwisely, in my opinion) references this research in chapter 8 of his "Dark Mission" book.

THEMIS Camera Images the Face in 2002

In chapter 11 of his "Dark Mission" book, Richard Hoagland goes into considerable detail about analysis of a THEMIS infra-red image from the Mars Odyssey probe, which orbited the planet at that time. In his long discussion of the shenanigans regarding this THEMIS/Mars Odyssey images, Hoagland again makes an interesting observation, which I reference below.

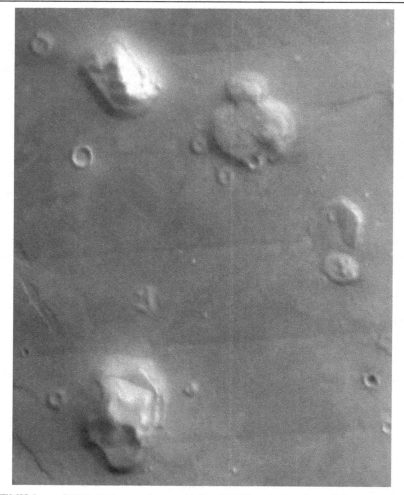

THEMIS Image V03814003 was taken on 24 Oct 2002[117] and, though much lower resolution than the MGS image, shows the face as being highly reflective (or emissive?) of infra-red radiation. Why is it so much different from the nearby features?

In 2003, Hoagland posted a detailed discussion of this image, and other related images, on his Enterprise Mission website[118].

Seeing the Face Even More Clearly?

Mars Express

In September and October 2006, ESA posted images of the face taken by Mars Express[119], along with a 3D-fly around video animation[120]. Some researchers criticised the 3D model as inaccurate

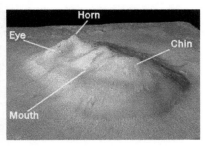

ESA Image of the "Face"[121] taken by Mars Express in 2006. Image ID 215131

The 3D ' Face' models by ESA[122] aren't consistent with each other[123]. Notice the lack of a "horn" on the image on the left. ID 208296

In the article describing the new images, the following text was included (my emphasis added):

> The array of nearby structures has been **interpreted by some <u>space enthusiasts</u> as artificial landscapes, such as potential pyramids and even a disintegrated city**. The idea that the planet might have once been home to intelligent beings has since inspired the imagination of many Mars fans, and has been expressed in numerous, more-or-less serious, newspaper articles as well as in science-fiction literature and on many Web pages.

> Despite all this, **the formal scientific interpretation has never changed**: the face remains a figment of human imagination in a heavily eroded surface.

> It took until April 1998, and confirmation with additional data from the Mars Orbiter Camera on NASA's Mars Global Surveyor, **before popular speculation waned**. More data from the same orbiter in 2001 further confirmed this conclusion.

The page also includes a quote from Carl Sagan:

> *"Imagination will often carry us to worlds that never were. But without it we go nowhere."*

The pattern is the same. Those who study and present evidence of possible artificial structures in Cydonia are relegated to being "space enthusiasts" – even though Dr Mark Carlotto, Dr John Brandenburg, Dr Tom Van Flandern, Prof Horace Crater and Dr Brian O'Leary are all scientists… Can we count this as a lie by ESA? I think we can!

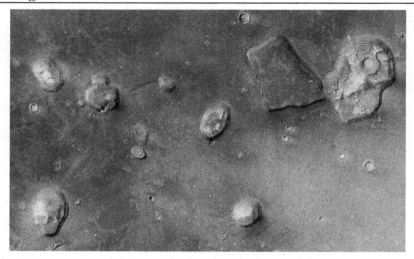

ESA Image of part of Cydonia region including the Face (top left) and an interesting structure over to the top right – which some have called the Bastion[124], taken by Mars Express in 2006[125] (cropped from full version).

Mars Reconnaissance Orbiter - HiRISE in 2007

On 12 August 2005, Mars Reconnaissance Orbiter was launched[20] and it contained the HiRISE camera, built by Arizona University. Very high-resolution images of the surface – giving resolution as good as 1 meter per pixel. (The original Viking 1 and 2 images were about 43m per pixel in resolution.)

On 11 April 2007, new images of the Face and Cydonia were released and posted on the HiRISE website[126].

In a posting entitled "Buttes and Knobs in Cydonia Region" from 11 June 2008, Alix Davatzes writes:

> The "face" has subsequently been imaged by many orbiters, including HiRISE (PSP_003234_2210), showing that it is simply a rocky mound, and the face-like appearance was due to a trick of shadows.

In 2010, I made a video of a "flyover" of the Face, using imagery in Google Earth from the HiRISE camera. Whilst the posting on the University of Arizona website claims it is a natural landform, we are still faced with the same issues as we were in 2001 – when the MGS images were released. We can see more detail in the image – but the overall shape – and features – remain. I would argue, as I think Mark Carlotto would, that we can just see the details of the erosion better. For example, at one location, we can even see a line of rocks that seem to have broken off together.

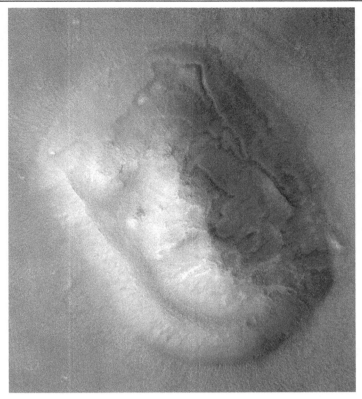

"Popular Landform in Cydonia Region"[127] – Newest "Face" image from April 2007 – Looks little different to the 2001 image, at this resolution.

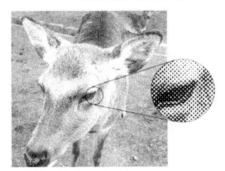

Left: One can draw an analogy of zooming in to a photo of a face in a newspaper or magazine. As one zooms in, the view of the face is gradually replaced with a matrix of dots.

Maybe none of us are real – because we are just a collection of dots (atoms) when seen close up?

Example section image from the surface of "The Face" demonstrating the high resolution of the image from MRO.

In the last chapter of his "Dark Mission" book, Richard Hoagland argues that the newer MRO images show the remains of structural elements of the Face's construction – and he feels he can see rectilinear geometry within the image. I find this difficult to see, myself. Perhaps Mark Carlotto's fractal image analysis technique, or some version of this, could be usefully applied to the high-resolution MRO images to see what is shown.

Conclusions

Overall, this is yet another area of research where the alternative version of the acronym "NASA" seems appropriate – "Never A Straight Answer." I do think NASA is hiding something about the Face and Cydonia and it is something important. Their behaviour and reaction towards those who have done investigations strongly indicates that someone knows a lot more than they will reveal. We will revisit these themes later in this book.

5. Pyramids of Mars

Dr Who fans will recognise this chapter's title from the fabulous Tom Baker story from 1975[128] – the year before the Viking mission arrived at Mars. Are there really some pyramids on Mars?

D & M Pyramid

There are other unusual features in Cydonia, close to the "Face". Perhaps the "second most famous" of these is the D & M "pyramid," which has been named after the two people that first studied it. In 1979, when Vincent Dipietro and Greg Molinar re-discovered the first Viking image of the face, they also noticed a strange pyramid-shaped landform. We can see its location at the bottom of the image, shown below.

This is a contrast-enhanced image of features in the Cydonia region, from the MSSS Website.[129]

In meetings of the Mars Investigation team in the 1980s, Richard C Hoagland argued this was a 5-sided shape, whilst a physicist from SRI, Lambert Dolphin (who had studied "lost cities" in Egypt) argued it was a 4-sided pyramid with a badly damaged face.

Erol Torun studied the Pyramid in some detail and in an article he wrote in 1988 or 1989, he describes it thus[130]:

> *The front of the D&M Pyramid (closest to the face) is formed by two congruent angles, with two larger congruent angles forming the sides. A fifth angle forms the rear section. The pyramid exhibits some domed uplift on its right side, and what appears to be an unusually deep impact crater further to the same side.*

> *The geometric regularity of the D&M Pyramid, together with its alignment with other enigmatic landforms, has led some to speculate that the object may have an artificial origin…*

He continues:

> *There is a theory that the northern Martian basin called Acidalia Planitia was once a shallow sea. This would place the area of Cydonia Mensae under study near the former shoreline. Small craters in this area appear to have been modified by water erosion, perhaps by shallow wave action. This would match the observations of recent researchers that linear features in this area may be lacustrine deposits resulting from shallow wave action at the edge of an ancient sea…*

Dr Mark Carlotto also discusses this in parts of his "Cydonia Controversy" book. Below, I include a higher resolution version from MGS image MOC2-484[131]. You will notice part of this image is in lower resolution than other parts, due to it being taken at a different time and then composited with other image strips (see the MGS MOC2-484[131] image page for more details).

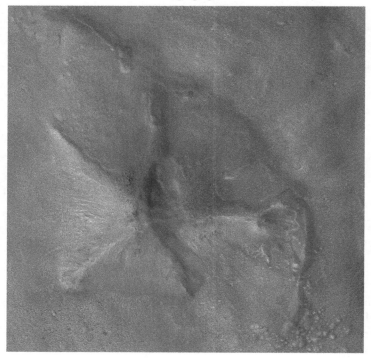

D & M Pyramid – MGS Image from 15 September 2003[132]

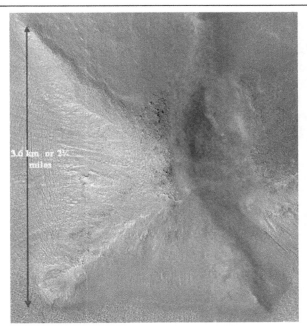

Left: From the information given with the image, it is possible to calculate the length of the sides of the pyramid. As Mark Carlotto and others have observed, the "Pyramid" appears to be aligned to the Martian North/South meridian, along the measurement line I have added, above.

For those that would like an even closer look at the D & M Pyramid, I recommend trying out the GigaPixel image viewer[133].

Another apparent Pyramid that is not that far from Cydonia is the "Starfish" pyramid. It sits next to another unusual feature called "The Fort."

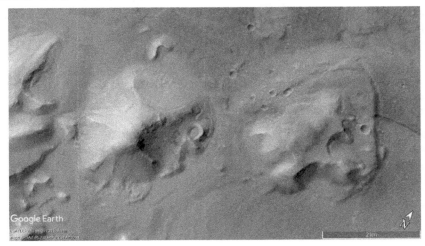

On the left[134], we have a structure which exhibits linear divisions in its shape. The landform on the right[135] is also unusual in character – hence dubbed "the Fort"[136].

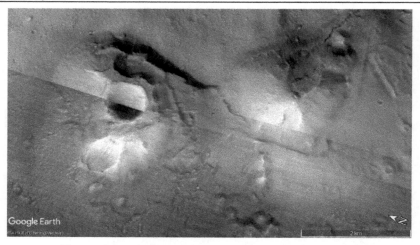

Unusual Bowl-shaped feature (left) at 39.3°N, 10.75°W

Links to high resolution MGS images of some of these features can be found on the MSSS website[129]. The images you will find there are not "contrast-enhanced" so they look grey and featureless. Most image editing applications such as GIMP[137], Photoshop, Paintshop Pro and Picasa [138]allow you to enhance the contrast to see the features.

Elysium Pyramids

There is a useful page on Holger Isenberg's site [139]which collects together some detailed information about these features. The Elysium Pyramids were actually pointed out by none other than Carl Sagan, in chapter 5 of his 1980 book "Cosmos"[140]:

> *The largest are 3 kilometers across at the base, and 1 kilometer high - much larger than the pyramids of Sumer, Egypt or Mexico on Earth. They seem eroded and ancient, and are, perhaps, only small mountains, sandblasted for ages. But they warrant, I think, a careful look.*

Sagan had seen them on Mariner 9 images, such as image number 4296-23, shown below. How ironic, then, that he would be so resistant to the studies presented to him by the Mars Investigation team later in the 1980s. We will learn more about Dr Carl Sagan in chapter 23.

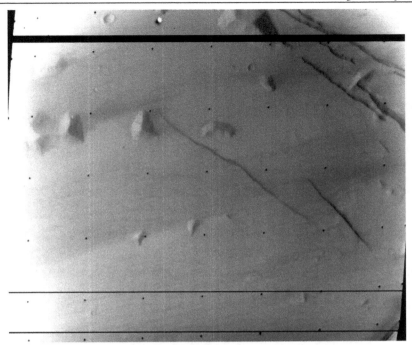

Mariner 9 Image of Elysium Pyramids at 15.5°N 198°W.

The images on Isenberg's page are now difficult to find in NASA's own archives. Fortunately, Isenberg has archived a copy.

Using "Google Mars," I found the area and grabbed the image below:

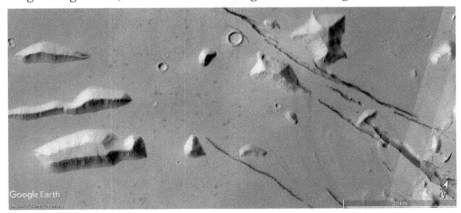

Google Mars Composite image of Elysium Pyramids at 15.5°N 198°W.

Just to the north of this view, I zoomed in on another portion of this landscape, which is shown on image P03_002318_1961_XN_16N198W from MRO.

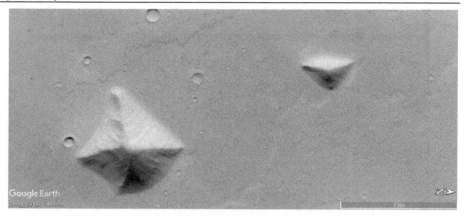

Google Mars image of Cerebrus Region Approx. 16.6°N 161,5°E, as seen on an MRO image.[141]

Close up inspection of these highly symmetric landforms reveals quite a few small craters, suggesting they are ancient. Also in this region, Isenberg notes another feature he likens to the Hebrew Letter "Nun," though to me it looks more like a sickle shape (although it has an "extra piece" at the base of the "handle," it would appear.

Originally from image FHA-01651, is this some kind of oxbow lake[142] on Mars, or is it an artificial structure, perhaps?

Another Elysium Pyramid?

A video by Walt Hain[143] led me to THEMIS image I03502047:[144]

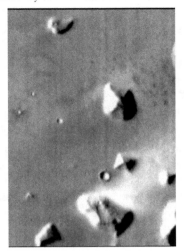 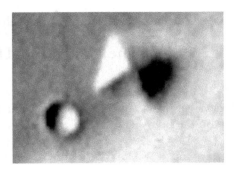

A portion of THEMIS image I03502047, zoomed on the right[144]

This structure is only about 110km South of the Elysium Pyramids. It seems to have a square base. There doesn't seem to be a higher resolution version of it available.

Conclusions

Of the anomalies shown in this chapter, the D & M pyramid has probably been studied the most. If I was pushed, I'd say that some of these pyramidal structures are, indeed, artificial – and there is some kind of link between these structures and the pyramids on Earth. In the next chapter, this link may become a little stronger.

6. More Martian Mysteries

In this chapter, we will inspect a few more apparently anomalous Mars images. Some of these are also discussed on an interesting web page by David Pratt[145]. Outside of the analyses done by a few scientists of Cydonia, I simply have not come across any detailed analysis of most of the other images in the rest of this book, so I cannot reference any such analyses!

The Tholus, The Crater and The Cliff

Tholus

Mark Carlotto also notes this feature, which again we will mention later. It is seen on the right-hand side of the 070A13 image (the second Viking image of the Face). It also seen on an image numbered 035A74[146] (taken just after the first face image – 035A72 – but I could not find this on any of the NASA image websites)

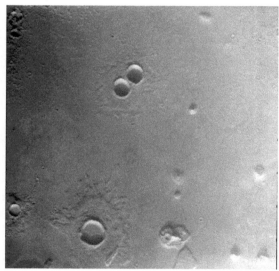

Image 035A74. Tholus is right of centre, below the middle of the image.

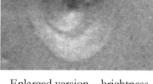

Enlarged version – brightness and contrast-enhanced.

About this Tholus, Dr Mark Carlotto writes (my emphasis):

> *The Tholus is a large mound-like feature, **not unlike Silbury Hill** in England, located about 30 km southeast of the Face. It was originally discovered by Hoagland in seeking additional objects from which to derive spatial and angular relationships. According to geologists, James Erjavec and Ronald Nicks: "They [the Tholus and several other features with similar morphologies] display no evidence of vulcanism, are not impact related and are difficult to explain as remnants of larger non-distinct landforms because of their symmetrical shapes and uniform low gradients."*

Once again, Mars Global Surveyor has now photographed this feature in higher resolution.

From Google Mars - MGS image of Tholus in Cydonia – from an original unprocessed source image on MSSS[147].

Dr Mark Carlotto again used his "shape from shading" algorithm to make a 3D model of this feature:

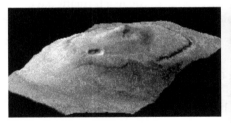 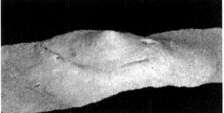

Two frames from a 3D-fly-around rendering by Mark Carlotto[148]

As with other features in Cydonia, it would be difficult to come up with a coherent geological / natural explanation for the formation of this feature.

Crater and Cliff

A few miles north of the Tholus, we find 2 adjacent features – called "The Cliff" and the Crater,

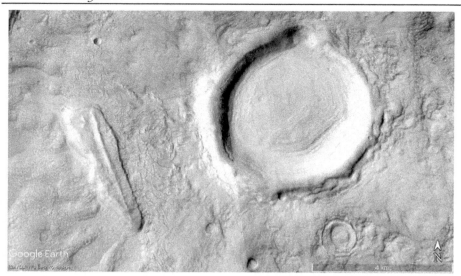

From Google Mars - MGS image of the Cliff (left) and crater in Cydonia – from an original unprocessed source image on MSSS[149].

This feature was first discussed, I think, by Richard C Hoagland, during his investigations of the Cydonia region. We can note, as he and others did back in the 1980s, the linear ridge which runs most of the length of this feature – a distance of some 2.3 miles or about 3.7 kilometers. The crater is approximately 3 miles/ 5.5 km wide and judging by the base looks to me like it might have been filled with water or some other liquid at one time.

Cydonia and Avebury?

Having looked at the Tholus, Crater and Cliff, we can now examine what appears to be a very peculiar coincidence indeed. This is not mentioned in Dr Mark Carlotto's "Cydonia Controversy" book, nor in Hoagland's "Monuments of Mars", although Hoagland has mentioned it on his "Enterprise Mission" website[150]. I am not sure which other researchers, apart from David Percy who discovered this strange "coincidence"[124], have discussed this. The correspondence is in the ratio of 7:1. Avebury is a large stone circle complex in Wiltshire, UK[151]. As I have said in presentations I have given over the years, I do think this incredible correlation indicates that there is a story here we're not being told. Someone knows more about this story than you or me – but they won't tell us. We will see more examples of this later.

The Tholus, Cliff and Crater image from Google Mars on the left, and an aerial view of Avebury from Google Earth on the Right. Using outlines, we will then attempt to overlay these, with some rotation.

Here, we can note two other apparent points of correspondence. Firstly, the tree line to the west of the village seems to match the western edge of the cliff quite well. Secondly, the contour line on the map curves with some correspondence to the "notch" in the rim of the crater in Cydonia

We do seem to have a very strange set of correspondences here – on a scale of 7:1 (Cydonia : Avebury). David Percy argues that there are many other points of correspondence and alignment[124]. I think I am in fairly close agreement with his statements:

> *We must stress again the impossibility, first, of mound plus crater rim in Cydonia and mound plus earth rampart at Avebury having by chance the same relative size, and being the same relative distance from each other... We could deduce from these findings that at some time in the past there has been physical and/or mental communication between Mars and Earth.*
>
> *NASA went to Mars on behalf of us all to look for life. In responding to the images of certain Cydonia features, is it not regrettable that the agency appears to ignore, and has attempted to debunk, evidence that strongly indicates the work of intelligent life?*

Nineteen Point Four Seven Degrees…?

Also, we can note another point from Percy's webpage – which I think was first pointed out by Richard C Hoagland:

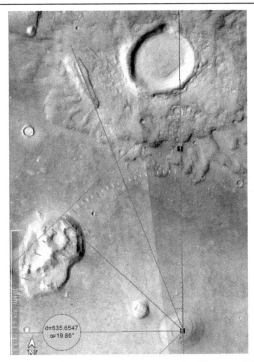

d=635,6547
α=19.86°

Cliff/Crater/Tholus arrangement – a significant angle?

We will take a "sideways step" here, for a few minutes and note that Richard C Hoagland has pointed out the apparent significance of the angle 19.47° and for example notes a possible correspondence in the image above. When I measured this using a screen protractor, I got a slightly different result. It also depends on where you draw the vertical line, of course. I could draw it to the edge of the "cut-out/bite" on the outside of the crater – or the edge of the "cut-out/bite" on the *inside* of the crater. Using the original lower resolution Viking image, as Percy (and Hoagland) did, this arrangement is more questionable.

However, we can observe that the line along the Cliff does *appear* to point directly to the centre of the Tholus, which is odd...

Hoagland and others argue that within spherical bodies, there is some type of flow of energy – "hyperdimensional" in nature. He also suggests it is a manifestation of "torsion fields," perhaps. Although I have reservations about Hoagland, there is no denying some of his observations here are correct. Also, in Chapter 2 of the "Dark Mission" book, Hoagland and Bara cover some very interesting territory related to this torsion physics and suggest it may tie in to what was discovered by Wilhelm Reich, for example. About a dozen other scientists were also investigating different forms of energy at about the same time. Reading Hoagland's "Hyperdimensional Physics" chapter, it did remind me of some things that Wilbert Smith said in some of his articles, presentations and in his unfinished book "The New Science."[152] Smith was also talking about "fields" of different types and "spin." Smith also talked about creating a special type of coil that he built, under the direction of "the boys topside" which he used to create a six-dimensional radio wave – which Smith said was a form of tensor energy[153].

One of the bases of the Hyperdimensional Energy or Torsion field theory involves consideration of the positioning of a tetrahedron within a sphere.

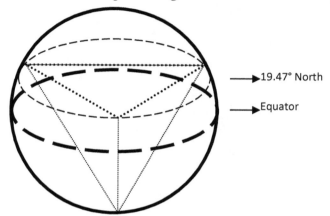

Imagine a Tetrahedron – A Pyramid with an equilateral triangular base – inside a sphere… This shows how the 19.47° angle comes about. If we have 2 tetrahedrons, we would have a 19.47° line marked out in each hemisphere.

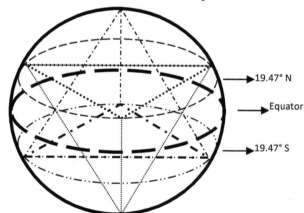

In his writings, Hoagland notes that the tetrahedron is the simplest 3D shape you can have – with a total of only 4 vertices. It has been suggested that this is perhaps where the Star of David symbol comes from, i.e. it was created by those who have a knowledge of this hyperdimensional energy.

It is quite common for this symbol to be seen inside a circle, too.

Hoagland suggests that this "energy" emanates from these points on a sphere and there is some evidence that this is true:

- Olympus Mons, 27 km high volcano on Mars – latitude 19 degrees

- Solar Maximum – most sunspots occur at latitudes of 19.5 degrees

- Red Spot on Jupiter 19.5 degrees.

- Big Island of Hawaii – latitude 19 degrees

- Dark spot on Neptune – latitude 19 degrees

- Alpha & Beta Regio – Venusian volcanoes – latitude – 19.5 degrees.

- Strongest El Nino currents occur on latitude – 19 degrees.

Hoagland also claims that several alignments of features on Mars suggest that those who built them knew about this energy, for example.

In chapter 2 of the Hoagland/Bara book "Dark Mission", there is some very interesting material about how hyperdimensional theory explains certain unusual facts about the Solar System – such as the distribution of angular momentum across the various bodies of the Solar System. The chapter points out that the Sun holds about 1 percent of the angular momentum, but about 99% of the mass in the Solar System. Jupiter has 60%[154] of the angular momentum. Though this is somewhat difficult to explain, it is argued that the distribution is the result of the fact that the Sun exhibits coronal mass ejections fairly regularly, and that it's possible that during the formation of the Solar System, larger ejections transferred more of the angular momentum to the planets themselves[155]:

> *The origins of the Solar System, the Sun also produced, much more intense ejections of coronal mass and it is these that would have transferred its angular momentum to the young objects in the cloud formation. The planets and the Sun formed at the same time, there is 4.5 billion years, from proto solar nebula.*

Following his analysis of Cydonia, with Erol Torun, Hoagland claims to have predicted, in the 1980s, that there would be a large storm system in the atmosphere of Neptune, which Voyager was about to photograph at close range. This turned out to be correct – it was also in the Southern Hemisphere of the planet, at roughly 19.5° latitude. Hoagland also made correct predictions about Neptune's magnetic field.

Hoagland/Bara may well be "onto something" in pointing out the relevance of rotation, to anomalous physical effects that are seen in various scenarios. For example, Bruce De Palma's simple "high speed rotating ball bearing experiments"[156] demonstrate anomalous effects on gravity – and this seems quite similar to what Canadian Radio engineer Wilbert B Smith also talked about in the 1950s[157].

Voyager image of Neptune[158]

Some kind of equivalent of hyperdimensions was considered as part of James Clark Maxwell's original equations in the 19th Century, but was discarded with the simplification of those equations by Oliver Heaviside and William Gibbs[159].

Carl Sagan enters the picture again here, in that in his 1980 "Cosmos" TV series, he made a presentation about hypercubes and "flatland."[160]

Carl Sagan, from "Cosmos" (1980) explaining a Tesseract.

In his presentation, Sagan eloquently introduces the concept of the fourth dimension, by considering how two-dimensional creatures ("flatlanders") would be unable to perceive three-dimensional space. He then points out that we, as three-dimensional creatures, would not be able to perceive the fourth dimension. He does not explore the concept any further than the illustration of the tesseract, however.

Sulphate Mountain

This striking image is seen in an ESA/Mars Express image entitled "Close-up perspective view of the 'sulphate' mountain - looking east," Released 19/01/2006 10:17 am[161]. The information about this image given on the relevant ESA webpage includes the following paragraphs[162]:

> *The depression of Juventae Chasma, located north of Valles Marineris, cuts more than 5000 metres into the plains of Lunae Planum. The floor of Juventae Chasma is partly covered by dunes.*
>
> *In the north-eastern part of the scene, there is a mountain composed of bright, layered material. This mountain is approximately 2500 metres high, 59 kilometres long and up to 23 kilometres wide.*
>
> *The OMEGA spectrometer on board Mars Express discovered sulphate on the surface of Mars and confirmed that this mountain is indeed composed of sulphate deposits.*

I haven't seen any other features like this on Mars. Why would the deposits build up like this? If anyone knows more about this, please write to me!

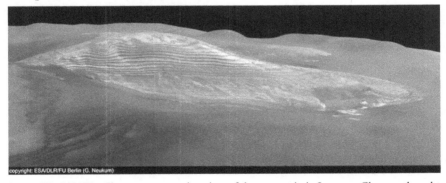

Image ID: 215277 - Close-up perspective view of the mountain in Juventae Chasma, thought to be composed of sulphates, looking east. The HRSC obtained this image on 26 March 2004 during orbit 243 with a ground resolution of approximately 23.4 metres per pixel. The scene shows the region of Lunae Planum, at approximately 5° South and 297° East.

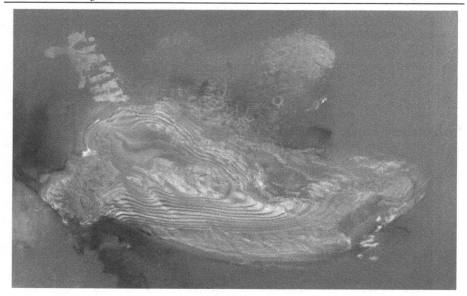

This is another view is contained in Image 054co01JuventaeChasma_H.jpg[163]

The same feature is seen in a 3D rendering with reference 0563D101JuventaeChasma_H.jpg[164]

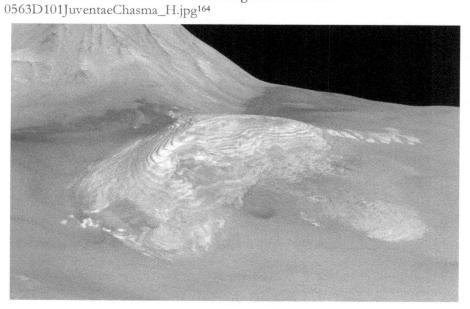

Another view of the terraced feature.

Whilst the 3D rendering shows this as a "convex" feature, the spacing of the contours or terraces or stripes is rather reminiscent of a mine.

Earth – Barrick Goldstrike Mine, North America[165]

Kalgoorlie Gold Mine (Australia)[166] (Google Earth Image)

Also notice the similarity of areas next to the main "works" or "terracing" in each of these images:

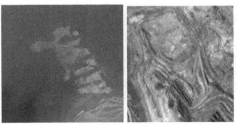

A Kofun, an Exclamation Mark or a Natural Feature?

Image ESP_020794_1860 from MRO contains another unusual feature[167]. The page for this image (which isn't for some strange reason, visible in the banner image – more on this later…) carries the following description:

> *Turn this image sideways (so North is to the right) and the high standing landforms look like an exclamation mark. The origin of these hills may be difficult to understand on such ancient terrain. The straight edges suggest fractures related to faults. Maybe this feature was lifted up by the faulting, maybe the surrounding terrain has been eroded down over billions of years, or both.*

What is this strange feature?[168]

It looks a little like a Kofun Era Tomb in Japan…[169]

The 'T' Channel

This was an image I became aware of in 2004, after visiting Tom Van Flandern's (now defunct) Metaresearch Site and reviewing his presentation about Mars Anomalies from 2001.

I have added an indication of the size of the feature. Though due to the lighting, this may look like a raised structure, it is actually a channel. Natural channels rarely end in a perpendicular 'T'-shape for obvious reasons. There is no obvious source or sink for any fluid that created this channel. I could find no other references to this image which attempted to explain its origin. Indeed, when I wrote to geologists in 2004, none of them chose to try and explain it.

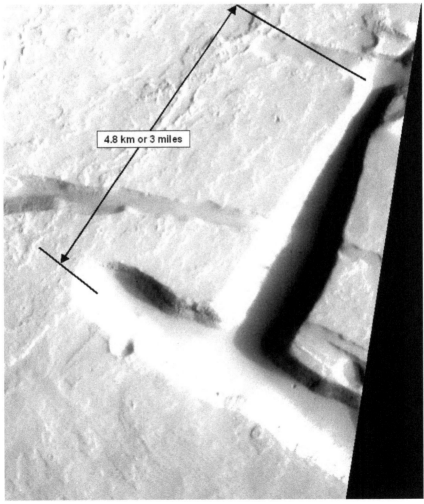

4.8 km or 3 miles

MGS Image SP243004 (South east of Olympus Mons)[170]

A Big Hint of a Rectangle?

This is covered in MJ Craig's Book "Secret Mars - The Alien Connection"[171] – is it a "giant lost city"? It can be seen in an image on the USGS website - e100046[172].

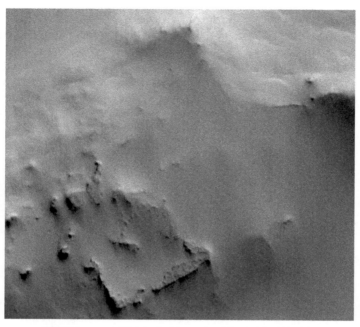

USGS Image e100046 – is this a ruined, buried building?[173]

At the bottom of this same image, I was given to wonder if we can see another partly buried, ruined structure...

This is at the bottom of USGS Image e100046 another corner of a ruined, building?[173]

Version without mark-up

Inca City

This has been shown and discussed on quite a few websites as evidence of artificiality. Even the MSSS website states the features are unexplained:

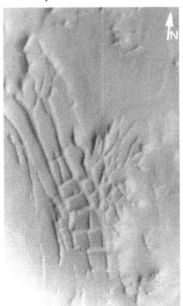

"'Inca City' is the informal name given by Mariner 9 scientists in 1972 to a set of intersecting, rectilinear ridges that are located among the layered materials of the south polar region of Mars. Their origin has never been understood; most investigators thought they might be sand dunes, either modern dunes or, more likely, dunes that were buried, hardened, then exhumed."

Mariner 9 image DAS 8044333
Dubbed the Inca City[174]

A new MGS image of the feature was released on 08 August 2002:

The Mars Global Surveyor (MGS) Mars Orbiter Camera (MOC) has provided new information about the "Inca City" ridges, though the camera's images still do not solve the mystery. The new information comes in the form of a MOC red wide-angle context frame taken in mid-southern spring, shown above left and above right. The original Mariner 9 view of the ridges is seen at the center. The MOC image shows that the "Inca City" ridges, located at 82°S, 67°W, are part of a larger circular structure that is about 86 km (53 mi) across. It is possible that this pattern reflects an origin related to an ancient, eroded meteor impact crater that was filled-in, buried, then partially exhumed. In this case, the ridges might be the remains of filled-in fractures in the bedrock into which the crater formed, or filled-in cracks within the material that filled the crater. Or both explanations could be wrong. While the new MOC image shows that "Inca City" has a larger context as part of a circular form, it does not reveal the exact origin of these striking and unusual Martian landforms.

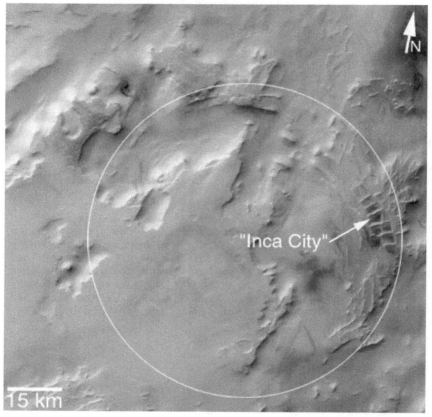

MGS Image MOC2-319[174] – This is their annotation – i.e. the indication of arcs or a circle in the structure.

Tubular Dunes?

This MGS image – number M04-00291[175] - has also been shown on many websites since it was released. Tom Van Flandern suggested that what we were seeing were long glass tubes! In the May 2001 press conference we mentioned in chapter 4, Van Flandern showed an image and stated:

They are networked - we've established that it is not an optical illusion. They are really tube-shaped. They're not easily explained as dunes or lava tubes. They seem to be translucent. In fact, in this case, that bright spot seems to be a specular reflection of the Sun, implying that the surface must be glassy or metallic. Natural surfaces don't ordinarily give you specular reflections.

However, Dr Mark Carlotto suggests that they are indeed dunes – in unusual relief, with unusual lighting. The bright spot seen roughly in the middle of the picture about one fifth of the way down is puzzling to me, however.

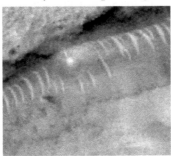

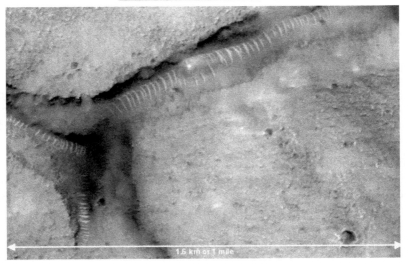

MGS Image M04-00291 – What does it show?[176] (Rotated)

One of the discussions about this object that I came across was with "Bad Astronomer" (yes, really) Phil Plait.[177] In an article dated 17 March 2004 on CNN's website, he offered his opinions when Richard C Hoagland had referenced this (and many other anomalous images). Here is a portion of what he said[178] (my emphasis added):

Phil Plait, author of Bad Astronomy (Wiley & Sons, 2002), which debunks space myths and common factual misconceptions, had for years not countered Hoagland directly, because he did not want to give a man he calls a "pseudoscientist" the "air time that he so desperately seeks."

> *Plait has two words for the latest claims of alien objects on Mars. The first is "garbage." The second and more scientific word is "pareidolia." This is the same phenomenon that makes us see animals or other familiar objects in clouds. "It's pretty common," Plait said of pareidolia. "Just a few months ago, a water spot on my shower curtain took on the uncanny form of the face of Vladimir Lenin." Plait took a picture of the liquid Lenin and uses it to illustrate his contention that, though objects on the surface of Mars can sometimes take on interesting shapes, they are **just a bunch of rocks**.*
>
> *"Hoagland's claims irritate me because he is promoting uncritical thinking," Plait told SPACE.com. "He doesn't want you to think about what you're seeing. He's trying to bamboozle you into believing what he's saying."*

Although this article is over 13 years old, I don't see the level of discourse has really changed all that much.

There is an obvious point here that doesn't seem to have been raised. We are not looking at astronomy – we are looking at geology – or even archaeology. Hence, Plait's opinions are largely irrelevant. Again, he toes the "light and shadow" line and doesn't offer any evidence to show that they are similar to other geological formations – he just offers a ridiculous, irrelevant comparison to something that happened when he was having a shower...

We will again see the pareidolia meme raised in chapter 7.

Conclusions

Collectively, the features shown in this chapter either show the results of some very unusual and very poorly understood geological processes, or they show the results of intelligence, construction and then decay. Either way, they should be the subject of vigorous investigation, not ridicule and debunking.

The most compelling evidence that I have seen regarding a possible link between Earth and Mars civilisations is the correspondence between Cydonia and Avebury. A statistician might argue that if you were to take a selection of three features from all the surface features on Mars and a selection of three features from all those on the earth, you would be bound to find a match somewhere – if you chose enough different groups of three in each case. However, considering the study which has been done of Avebury regarding its history, which likely started over 100 years ago, and the completely separate study of Cydonia which didn't start until the 1980s, it is far too much of a coincidence that these two places have matching architectural (there, I said it) features.

7. "Welcome to the Pleasure Dome"

In about 2004, when I had started to re-investigate the UFO/ET topic, one of the most striking images that I came across was one of some kind of "dome" structure on Mars. I think I first came across this on Jason Martell's "X-Facts" website[179], but I am not sure. It was apparently discovered independently by four different researchers - in 2001; Petranek (being the first), M. Tonnies, J. Danger, and Fitzgerald. Since then, I have seen little useful discussion of this image and only one considered analysis. This image is not featured in Dr Mark Carlotto's "Cydonia" book, it is not featured in Hoagland's "Monuments of Mars" book. It is not featured in MJ Craig's "Secret Mars" Book[180]. I have shown this image in my presentations many times since 2004 and no one has offered a sensible explanation for it. About the closest I have seen is that it is some type of "lava tube" formation.

The image number is m01501228 - taken on 19th May 2000[181]. I also came across the same image posted on Malin Space Science Systems website[182]. (This version unlike the version posted on the USGS website, is not contrast-enhanced.)

From the information on these pages, I calculated the crater to be roughly 530 meters across. There are 2 things which are clearly very curious about this image. Firstly, there appears to be a well-defined dome structure and secondly there appear to be "ribs" on this dome. Can this be a natural formation?

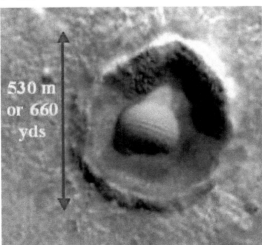

MGS Image M01501228[183] – What does it show? It's at 36°36'N and 27°29'W.

In reviewing the whole image strip in which this feature was found, we see quite a few other anomalies – which I don't recall seeing elsewhere… We can see what appear to be translucent domes and even an almost octagonal crater!

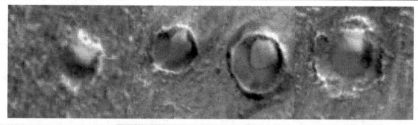

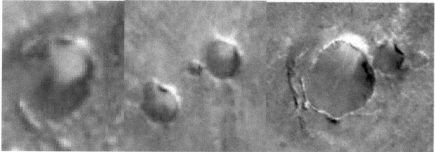

These are all "craters" from the M01501228 image strip – but they are enlarged by different amounts, so that they fit across this page. Do they have a clear, domed cover?

An answer to what the images show?

The best analysis I have seen of this dome image is on a web page entitled "Is This An Artificial Construct On Mars?"[184] It does not claim to prove that the dome is an artificial structure, but it does include some interesting additional images and some excellent commentary:

> *...the illumination or sun angle is from the upper right and what is seen in this image supports that contention, as well as raising a number of interesting questions. Both craters appear to have large spherical objects contained within their respective boundaries. On the left side (M1501228d) this sphere appears as if it is rising from beneath the surface of Mars like some gigantic egg or puffball. One can clearly see the crenulated edge of the crater's overhanging lip and the associated shadows cast by it.*

Wider field view of "Domesville"

Using Google Earth/Mars, I was eventually (after some searching, despite having the co-ordinates) able to find the field where the "dome" lies.

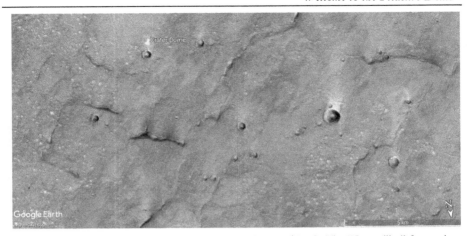

Above: Area in which dome lies, as shown in Google Mars/Earth. The "Caterpillar" feature is about 1 mile north of the dome, seen about half way down the image. In the Palermo Page, it is suggested that the "snake like" features might be tunnels.

More Domes?

I then found quite a few similar craters, including a virtual twin! I have included a few additional images below. I don't know which MGS/USGS image numbers these additional features are on, however. On Google Earth/Mars, the images don't seem quite as clear. Additionally, the newer HiRISE/MRO images (which we will discuss later) don't seem to be available there.

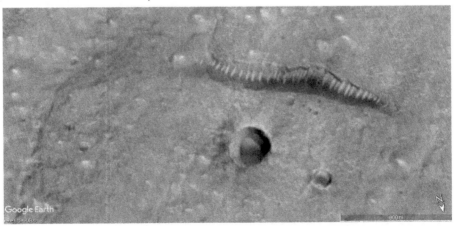

Odd craters (one with dome?) and similar caterpillar-like feature.

Odd craters, with domes?

More of the same…

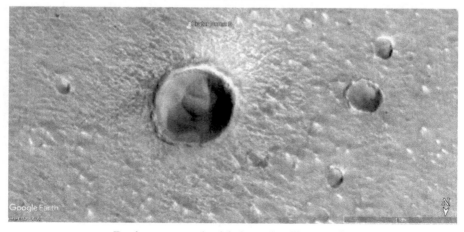

Do the craters on the right have glass-like covers?

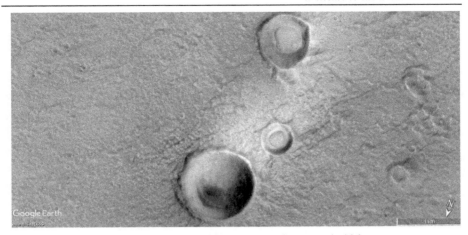

More strange craters with some type of structure inside?

Do we have a right-angled structure, north of the 2 craters?

More glassy-looking craters on the left, and another possible dome-in-crater...

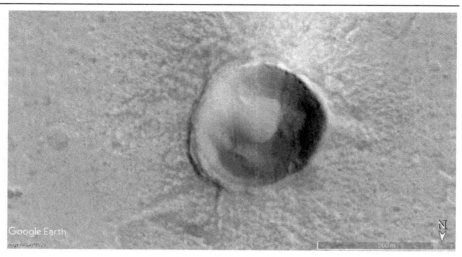

Is this a Crater Dome Twin?

The same question now arises regarding this dome-like structure in a crater. Are we looking at archaeology or geology?

When I saw the first crater dome image (the clearest), my immediate reaction was that it was a geodesic dome, like Disney's Epcot centre:

Epcot Centre – a Geodesic Dome[185]

While considering this idea and looking for more information about the crater dome structure, I came across a forum posting which drew a comparison to the dome which was constructed at the Amundsen-Scott South Pole Station back in 1975[186].

Amundsen-Scott Station Geodesic Dome - (image source: Eco-Photo Explorers)

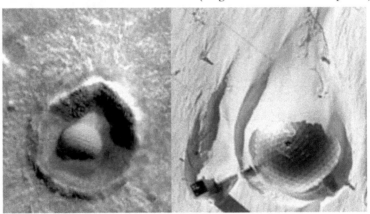

Mars Crater Dome (about 200m across) vs. Amundsen Scott dome 50m across.

It was the thread on the forum that led me to the new MRO HiRISE images, taken in 2008. These images have about seven times higher resolution than the earlier MGS images.

HiRISE Images of Crater Dome

On 10 February 2008, Mars Reconnaissance Orbiter (MRO) took new images of this strange feature and they were posted on a University of Arizona Webpage...

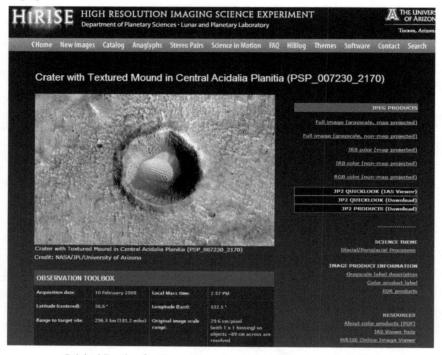

Original Posting for Crater Dome Image – or "Textured Mound"[187]

It seems then from this title, we can take this as a convex feature within a concave one. But why was there so little commentary? What is this feature? How was it formed...?

Using the links given with this image, we can download high resolution (and large) files and then "zoom in" considerably and obtain the views shown below. In the first enlargement, it seems quite clear to me that there is a straight looking "tube" protruding out from the dome like structure, going underground – in the direction of the slightly more "angular" feature on the crater rim. How very odd.

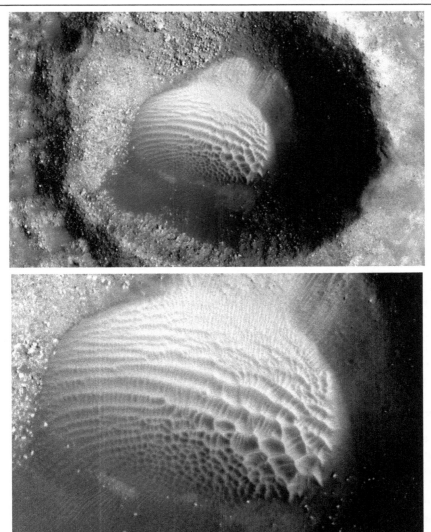

It still looks like a dome to me!

This is still, to me, one of the most arresting images that I have seen from the surface of the planetary bodies that have been photographed in my life time. You can therefore imagine my puzzlement when, a year or two after I first downloaded the high-resolution images, I revisited the University of Arizona's webpage to check details of the image. This is what appeared:

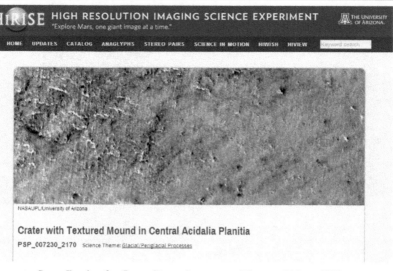

Later Posting for Crater Dome Image – or "Textured Mound"[188]

As you can see, they had removed the headline image – the most interesting one. They replaced it with a featureless view of part of the same area. Oh, sure – you can still find the "dome" – if you download JPG image[189] or one of the 2GB files that are linked at the bottom of the page! "Nothing to see here!" You agree, of course? Remember, the face was proven to be a collection of hills and we are just seeing patterns…

Explaining the Formation of the Dome

Over the years, quite a few people have alerted me to something I already knew. The single video that I have seen which specifically references the "dome" and attempts to explain its formation is called "Symbols of an Alien Sky"[190]. It is made by the "Electric Universe" group. Much as I admire their work and I agree with the explanations they have made for quite a few space-related observations, they seem to refuse to acknowledge certain things (such as the relevance of energy to how the WTC was destroyed). In the case of Mars features, I don't really think their explanation is a good fit. Let's have a look at the comparison they draw:

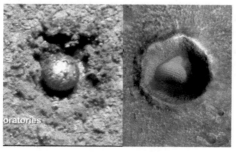

Side by side comparison of the plasma discharge experiment and the "dome."

I can see at least one significant difference – it is a feature we have already observed – the tube-like structure emanating from the dome. Also, the crater surrounding the dome is much more clearly defined than the loose material shown around the sphere of material formed during the discharge. Also, the surface of the dome has regular lines – which, admittedly, could be some type of frozen or solidified material – or some type of large crystal growth. The surface of the dome is observably different to the "plasma discharge" ball.

In their video, they overlook the other features we have already observed or examined – i.e. the "caterpillar" feature and snake like features etc.

I would therefore contend that the dome is an artificial structure and I would not be surprised at all if it was part of a larger system that is connected, as the "Palermo" web page mentioned earlier, proposes.

It appears no further information is forthcoming from NASA, JPL or the University of Arizona.

The "Caterpillar"

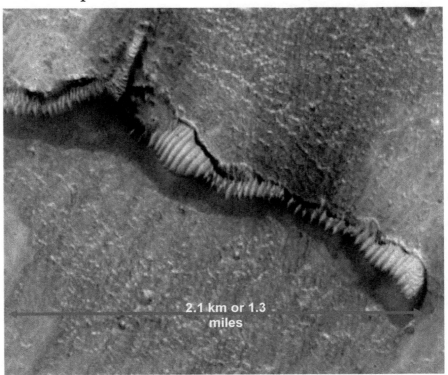

2.1 km or 1.3 miles

What is this feature, from strip M01501228[191]? It is about 1 mile north of the "dome". Is this really a collection of odd-shaped dune structures? I find it hard to believe.

I nicknamed this feature "The Caterpillar". Again, I've seen no explanation of what this structure is, or how it was formed.

A Closer View

In the same MRO image, numbered PSP_007230_2170[188], from 10 Feb 2008, we can again see, in high resolution, the strange appearance of this feature:

"Caterpillar" Close up.

Are these just sand dunes? Regularly spaced rock plates? How did they form?

Conclusions

I would argue that this "dome" represents one of the best candidates for being an artificial construct on Mars. There has been far less discussion of it than Cydonia and "the Face," yet the dome images, showing its symmetry and spherical nature seems, to me, much more obviously constructed. The surrounding anomalies seem, to me, to offer additional evidence that it is part of some kind of network of domes and tunnels.

8. Sojourning on Mars to Find Anomalies with Opportunity, Spirit and Curiosity

Richard C Hoagland and quite a few other people have been talking about anomalies seen on pictures from all the Mars Rovers since 1997. I have also been talking about them since about 2004. In November 2014, however, I was contacted by Richard D Hall[192], who had begun to consider the idea that the rovers were not actually on Mars. (This was due to him – and me – realising that the manned Apollo missions could not have landed on the moon.) Over the next 12-18 months, Hall gathered evidence to support this hypothesis, and some of this evidence is discussed in chapter 17. For the purpose of presenting these images in this chapter, then, we will just look at the anomalies, based on the assumption that the rovers are, indeed, on the red planet.

The Mars Rover Missions

There have, to date, been four successful NASA missions, as shown in the table below (data from NSSDCA Master Catalog[193]):

Name	Launch Date	Landed
Sojourner (Pathfinder)[194]	04 Dec 1996	04 July 1997
Spirit[18]	10 June 2003	04 Jan 2004
Opportunity[195]	08 July 2003	25 Jan 2004
Curiosity (Mars Science Laboratory)[22]	26 November 2011	6 Aug 2012

The Sojourner rover operated for nearly three months, but the other three have operated for many months – years even.

In the archives, there are now many thousands of images allegedly from the surface of Mars. (The Spirit Rover archive alone has over 128,000 images, as of Dec 2017.) These rover images arrived on earth when internet access had become much more widespread and therefore, unlike many of the earlier missions, new images were studied more promptly and in much more detail - by thousands of people. There are whole websites and YouTube channels devoted to analysing these images and pointing out anomalies. Here, I will only cover a few of the most interesting anomalies of which I have become aware.

Richard C Hoagland and Mike Bara have also highlighted some of these anomalies on Hoagland's "Enterprise Mission" website[196].

A Sphinx?

Of course, for anyone that thinks it unlikely that a huge face-like structure was built on Mars, the idea there might be a sphinx there would seem to be even more unlikely. Nevertheless, this is what Bara and Hoagland suggested they had found in a Pathfinder image…

A portion of Pathfinder/Sojourner image 80881[197] - Sol 2

Is this just a random pile of rocks? Or, with the hill that looks vaguely pyramid shaped, are the annotations below accurate?

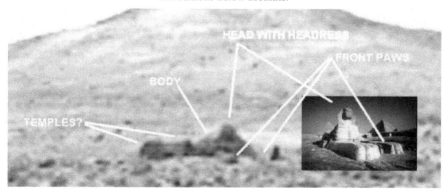

Annotations and inset by Hoagland/Bara[198]

Also, Hoagland claims the "Sphinx" is at 19.5°N and 33°W. Essentially, this is confirmed on the mission page, for the Pathfinder/ Sojourner rover[194]:

The landing site in the Ares Vallis region of Mars is at 19.33 N, 33.55 W.

Here is an MRO image of the landing site:

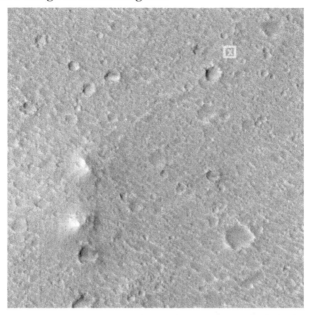

MRO Image of Pathfinder Landing Site[199]. The raised features were dubbed "Twin Peaks". They seem to be the only peaks for miles around. The yellow X marks the position of the Sojourner rover.

We cannot *obviously* see that the peaks have a square base, but looking at the lower of the two, can I see a sort of eroded square...? If you look at the landscape around the area, they seem slightly unusual...

How Should we "Handle" This Image?

An image, taken by the Curiosity Rover on 30 January 2013[200], contained a peculiar shiny object, which looks like a bicycle handlebar sticking out of a rock!

Here is a cropped, contrast-enhanced version

An article by Nancy Atkinson dated 23 Dec 2015 on "Universe Today" [201] contains an "explanation" for this artefact. Ronald Sletten from the University of Washington, who is a member of the Mars Science Laboratory team, was reported to have said it was a "completely natural feature."

> *"[On Earth, as on Mars] often you can see knobs or projections on surfaces eroded by the wind, particularly when a harder, less erodible rock is on top. The rock on top of the projection is likely more resistant to wind erosion and protects the underlying rock from being eroded. The shiny surface suggests that this rock has a fine grain and is relatively hard. Hard, fine grained rocks can be polished by the wind to form very smooth surfaces... while the surfaces not directly being eroded by wind may have a fine layer of reddish dust or rock-weathering rind. The sandblasted surfaces may reveal the inherent rock color and texture. This knob has a different type of rock on the end of the projection. This rock may vary in composition or the rock grain size may be smaller."*

Sletten then illuminates the reader regarding the phenomenon of "ventifacts" – wind-eroded surfaces – which one presumes he is saying that this is what the "handle" is an example of (though he doesn't explicitly say this).

Ventifacts formed from dolerite rock in Taylor Valley, Antarctica. This group of rocks likely started as a single rock that over time has broken apart. The wind has eroded the surfaces that appear gray, while the reddish part of the rock is simply the weathered surface or rock patina. The inset picture shows another example of a rock breaking apart in place and being eroded by the wind.

Rocks showing ventifact features.

Here we see a wind-eroded rock that has originated along a contact zone between granite and dolerite. The dark-colored dolerite is resistant to erosion and is polished smooth by the wind. The granite is also shaped by the wind; however, since it is coarser-grained, it is not polished as smoothly as the fine-grained dolerite.

Rocks showing ventifact features.

My response to this explanation is that the features shown look nothing like the object on Mars. It seems like a geologist is just using a fancy word to try and explain something he cannot explain.

A related question I will raise at this point is that the atmospheric density at the earth's surface is 1.217 kg/m³, according to a NASA website[202]. The same website gives[203] the same figure for Mars - ~0.02 kg/m³. My question is, with an atmosphere 60 times less dense, what effect does that have on the ability of dust to erode objects? How much more rarefied would the dust be? How can it be suspended in such a rarefied atmosphere? I know nothing... please write to me and explain.

The Metronome

The website Marsanomalies.com[204] held a peculiar image, shown below (the website is now not showing its images correctly). It is a Spirit Rover image from Sol 1402 - it was also imaged on Sol 1401 (hence, the images were taken on 12 and 13 December 2007). These images have just been cropped and contrast-enhanced – they have not been processed in any other way.

This image was taken by the Spirit Rover[205]. What is this object?
It looks like an old-fashioned mechanical metronome!

It seems as if the object may be attached to something else too, as suggested on a page giving a 3D analysis of the object[206]. The left-hand image was taken on Sol 1401[207] and the right hand image was taken on Sol 1402[205].

I am guessing that someone will suggest that this "metronome" object is another ventifact, and the broken off pieces which would "fit" with it have rolled away, down the slope?

Aerial Objects

Here are some photos of what appear to be solid objects in the Martian sky. NASA would normally suggest these are instances of dust on the camera or in the case of light phenomena, simply reflections of sunlight. Of course! What else could they be?

In this 2004 image from Spirit Rover[208], what is this object in the sky?

In the image below from Sol 665, what is the glowing light which appears to move quite some distance between 2 frames?

This is another image, taken around the same time[209]:

This Curiosity Rover image taken on 09 Aug 2012[210] is rather uninteresting… apart from the three faint objects in the sky which can be seen when the image is cropped and contrast-enhanced.

Are there meant to be any flying objects in the skies of Mars? Mars only has 2 satellites Phobos and Deimos, so if we assume we are seeing both the moon's in this image (which could be verified if necessary), we have at least one unidentified aerial object.

"Eddy the Eagle"

Another Curiosity Rover image, taken on 29 Jan 2015[211] looks somewhat uninteresting until you look at what appears to be some dirt on your screen...

A Square Block, or a Line-of-Sight Effect?

An image taken on Sol 581, by the Spirit Rover, which I came across some time after 2005[212], was originally posted on Zechariah Sitchin's Website[213]. It appears to show a squared off block…

A Dust-Devil in the distance – and can we see a squared off block? What about that rock on the middle left of the photo?

Are we seeing some kind of loop (reminiscent of the handle of a large jar or vase), or is it just a line-of-sight effect?

If you look at other Mars rover images, you may find others with hints of objects with machined/tooled features.

Don't Lose Your Head Over This Stuff!

Another image I came across on a website also contains a strange-looking rock – and indeed, it may be only that… It is probably just a line-of-sight effect. However, it is included here for you to go and check out yourself. It's from another Spirit Rover image, taken on Sol 513[214]:

From the bottom left quadrant of the image…

Another Face on Mars?

Again, perhaps we *are* in this case seeing a random placement, but the symmetry seems quite high – and the smaller rocks look "placed."

Conclusions

The most peculiar object seen in this chapter seems to be the "metronome" and I think this is something that must have been constructed and did not form naturally. The true nature of the other objects laying on the ground is less clear. The aerial objects are difficult to explain and indicate, in some cases, that either there are craft visiting Mars as well as the Earth, or the rovers are not on Mars at all.

9. The Waters of Mars

Dr Who fans will recognise this chapter's title from a David Tenant season special episode!

Most sources agree that there was once a lot of water on Mars. For example, the entry for Mars on Britannica.com[215] reads:

> *There are intriguing clues that billions of years ago Mars was even more Earth-like than today, with a denser, warmer atmosphere and much more water—rivers, lakes, flood channels, and perhaps oceans. By all indications Mars is now a sterile frozen desert, but close-up images of dark streaks on the slopes of some craters during Martian spring and summer suggest that at least small amounts of water may flow seasonally on the planet's surface and may still exist as a liquid in protected areas below the surface. The presence of water on Mars is considered a critical issue because life as it is presently understood cannot exist without water.*

I think this is an excellent summary, based on my own study of some of the Mars images and data. If you spend a few hours looking at the imagery on Google Mars, it's hard to ignore what seems to be clear evidence of dried-up riverbeds and other water-formed features.

We will discuss a few of these features in this chapter. However, we will also examine NASA's very peculiar attitude to Martian water. The reporting about the current presence of water on Mars, over the last 40 years, has been far too cautious, coy, ambivalent and even contradictory.

Water Ice Frost on Mars!

This was the description given to what was shown in an image from the Viking lander taken on 18 May 1979:

"This color image shows a thin layer of **water ice frost** on the Martian surface at Utopia Planitia."[216] Viking 2 Image - 21i093

If you're thinking this image looks a "bit too reddish-orange," we will look at this issue later.

Further questions about both the amount of water and the colour of the sky on Mars are raised by an ESA photo published on 28th July 2005.

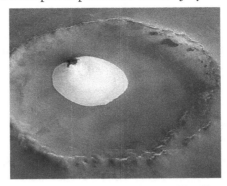

"Water Ice in Crater at Martian North Pole" – a Mars Express Image[217]

Why is the ice blue in colour? Could it be reflecting the colour of the sky on Mars? (Again, we will return to this issue later.)

So, both NASA and ESA have "told us" that there is Water Ice on Mars. But this, of course, doesn't mean we would have life… it's too cold there… everything is in "deep freeze" – or so we are told…

Liquid Water on Mars

In studying the Hoagland/Bara "Dark Mission" book, in chapter 9, we read:

> *The final proof of this came ironically not from Odyssey, but from the venerable old Mars Global Surveyor. Surveyor has carried an instrument that up until this point had been pretty much an afterthought, called the Thermal Emission Spectrometer (TES). One of the most stunning (and stunningly ignored) results from this instrument was its finding that during the summer on Mars (remember, Mars' year is about twice that of Earth's), the regions of Mars even above 40° latitude warm to a ground temperature of over 60° F. Obviously, this is well above the threshold at which water can exist in a liquid state, and resoundingly destroyed the last objection to the seeps as liquid water.*

Though I found it difficult to find the exact data on which this statement is based, an article dated 28 September 2017 on Phys.org[218] essentially says the same thing:

> *Preliminary weather reports from the Curiosity's Remote Environment Monitoring Station (REMS) are showing some surprisingly mild temperatures during the day. Average daytime air temperatures have reached a peak of 6 degrees Celsius at 2pm local time. A Martian day – known as a Sol – is slightly longer than Earth's at 24 hours and 39 minutes. Temperatures have risen above freezing during the day for more than half of the Martian Sols since REMS started recording data. Because Mars's atmosphere is much thinner than Earth's and its surface much drier, the effects of solar heating are much more pronounced. At night the air temperatures sink drastically, reaching a minimum of -70 degrees just before dawn.*

This article was quoted in another, dated 29 November 2017 on Space.com[219], which contains additional information about the higher-than expected surface temperatures.

However, the presence of liquid water on Mars is still moot, because, of course, the atmospheric pressure is so low that any liquid water would immediately evaporate, wouldn't it?

Mud on Mars!

I was intrigued to read this report from January 2004, on the BBC News website[220] (emphasis added):

> *Scientists are intrigued by the marks Spirit's airbags left on the surface. The soil shows an unusual cohesiveness, almost as if the soil grains were stuck together like mud. Jim Bell says that they see "scratch marks from where the airbags were retracted and there are places where rocks were actually dragged through the soil and the soil was kind of stripped up and folded in some places in very interesting and quite alien textures". Steve Squyres is also puzzled: "The way in which the surface has responded is bizarre. **It looks like mud, but it can't be mud**. We're going to have a real interesting time trying to figure this stuff out."*

Sand Ridges – seen in many rover photos…[221]

If you examine the image closely, you can see ridges in the sand. If you were to consider the impressions left by a child's toy (which had caterpillar tracks) on the dry, sandy portion of a beach, you may be given to wonder "how could the ridges seen in the photograph form in dry, dusty sand?"

However, after I had posted the above photograph, along with a few pages of other images included in this book, someone with the initials CLD wrote to me saying:

Dry dusty sand? No it's too cold. Steve Squyres is correct, it's not mud either. The temperatures are anywhere from a -170° to a -20°. If there was permanent water, these wheel tracks would not exist because the water/soil would be frozen solid. What you are seeing is soil adhesion at very cold temperatures. Suggest a trip to Antarctica to see similar behaviours of soil in extremely cold climates. This image is taken early on in the program when we didn't understand the dynamics of soil. We have since found many of the answers.

I don't think I will be taking a trip to Antarctica, just for the sake of verifying what CLD says… I think I will just "carry on" with the next image…

Other photos from the Spirit and Opportunity rovers also seem to show evidence of water *flowing* in the intervals between when photographs have been taken. The image below was originally found on Charles Shults III's old Xentotechresearch website (we will cover more of his research in chapter 11).

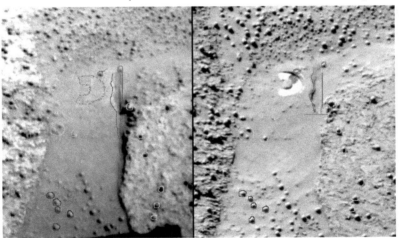

We can see the circular feature has disappeared between the left[222] and right [223]photographs. – and it is also shiny – from a thin layer of… liquid?

The semi-circular feature is from the rovers Mossbäuer (spectrometer) instrument[224], so how could this impression form in a dry dusty soil?

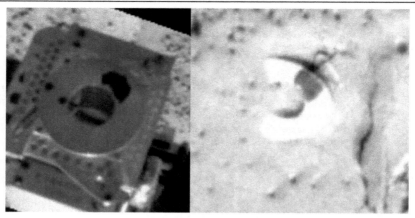

With this pair of images, Shults observed[225]: "The dark area on the Mossbäuer exactly matches the soil that was removed from the print. We see that this soil, which is very different from dry sand, has the ability to take a sharp print and stick to a surface pressed against it."

A Report in June 2000

On 26 June 2000, the website "Astronomy Picture of the Day" (APOD) featured this image[226]:

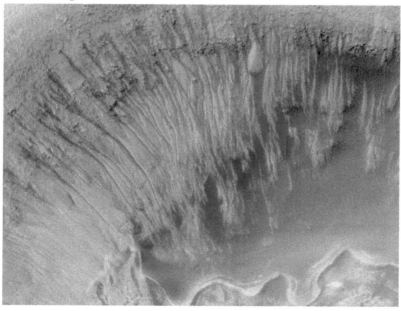

Explanation: What could have formed these unusual channels? Inside a small crater that lies inside the large Newton Crater on Mars, numerous narrow channels run from the top down to the crater floor. The above picture covers a region spanning about 3000 meters across. These and other gullies have been found on Mars in recent high-resolution pictures taken by the orbiting Mars Global Surveyor robot spacecraft. Similar channels on Earth

are formed by flowing water, but on Mars the temperature is normally too cold and the atmosphere too thin to sustain liquid water. Nevertheless, many scientists now hypothesize that liquid water did burst out here from underground Mars, eroded the gullies, and pooled at the bottom as it froze and evaporated. If so, life-sustaining ice and water might exist even today below the Martian surface -- water that could potentially support a human mission to Mars. Research into this exciting possibility is sure to continue!

A Report in 2006

After another 6 years had elapsed, we see a similar photo that was widely covered in the media at the time.[227] Why are they acting with such surprise? The first evidence of water was said to have been discovered 21 years earlier!

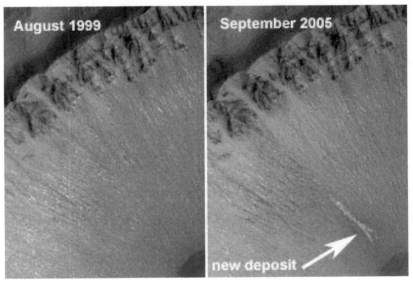

NASA photographs have revealed bright new deposits seen in two gullies on Mars that suggest water carried sediment through them sometime during the past seven years.

"These observations give the strongest evidence to date that water still flows occasionally on the surface of Mars," said Michael Meyer, lead scientist for NASA's Mars Exploration Program, Washington.

A Report in 2011

An animated image generated from MRO data appeared on 04 Aug 2011[228]:

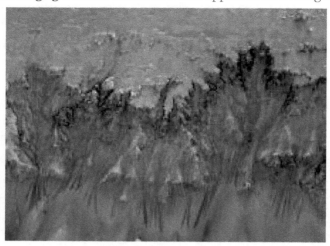

"NASA Spacecraft Data Suggest Water Flowing on Mars."

The report read:

> PASADENA, Calif. -- Observations from NASA's Mars Reconnaissance Orbiter
> have revealed possible flowing water during the warmest months on Mars.
>
> "NASA's Mars Exploration Program keeps bringing us closer to determining
> whether the Red Planet could harbor life in some form," NASA Administrator
> Charles Bolden said, "and it reaffirms Mars as an important future destination
> for human exploration."
>
> Dark, finger-like features appear and extend down some Martian slopes during
> late spring through summer, fade in winter, and return during the next spring.
> Repeated observations have tracked the seasonal changes in these recurring
> features on several steep slopes in the middle latitudes of Mars' southern
> hemisphere.

Again, this is yet more clear evidence of liquid water flowing. After a further
four years, however, it is still not openly acknowledged, it seems…

Report on 29 Sep 2015

A report on BBC News, with an accompanying 2-minute video, was posted on
29 September 2015 with the headline "Mars satellite hints at liquid water."[229]
The video description read:

> NASA scientists believe that dark stripes on Mars are caused by trickling water.
>
> Data from one of the US space agency's satellites shows the features, which
> appear on slopes, to be associated with salt deposits.

Such salts could alter the freezing and vaporisation points of water, keeping it in a fluid state long enough to move.

One of the images shown in the video

Captions in the video announce:

- *Mars has water…*

- *NASA's announcement overturns some red planet theories.*

- *"Mars is not the dry, arid planet we thought of in the past"*

- *"Today we are going to announce that under certain circumstances, liquid water has been found on Mars."*

- *"Could this water support life?"*

- *"Mars used to have a giant ocean"*

- *"Something has happened to Mars. It lost its water in the atmosphere and on the surface, for the most part, but we still have this question. Did life arise on Mars?"*

- *In other words… nobody knows yet.*

The video shows images from 2007 and 2011 – in other words, it seems to have taken them up to 8 years, or even 15 years or perhaps *even* 36 years to work out the situation with Martian water in frozen or liquid form. Is it just me?

Let us also note the clear evidence of surface moisture that NASA/JPL had from the images shown earlier – which were taken in May 2004. We also have mission controller Stephen Squyres comments about mud… Why is this such a problem for NASA?

The 2015 Film/Movie "The Martian"

It seems that the 2015 "we've found water on Mars" announcement was somehow related to the release of a new film starring Matt Damon. A story on Breibart.com read as follows[230]:

> **Ridley Scott Confirms NASA Timed Mars Water Find To Boost Damon's 'Martian' Movie**
>
> *On one level, The Martian may be functioning as a giant advertisement for Nasa, but the close collaboration between the space agency and Ridley Scott's film-making team has resulted in the director remaining blasé about the dramatic announcement of evidence of flowing water on Mars. "I knew that months ago," he said in response to the news. ...*
>
> *Scott said he had seen the photographs of water flows "about two months ago" – meaning that it was too late to incorporate the revelation into the film's narrative.*

The article also says:

> *Keep in mind that NASA's big announcement is NOT that water has been found on Mars. The news is that NASA has found only SIGNS of water on Mars.*

Which again adds a "flavour" of ambivalence to their statements. How incredible is it that nothing was mentioned of the discovery of signs of liquid water on Mars as far back as 2000?

Conclusions

The repeated reporting of the discovery of water on Mars over the last 40 years is indicative either of extremely sloppy journalism, or it is part of the cover up relating to knowledge about the presence and history of extra-terrestrial life in our Solar System. When I consider the other evidence presented in this book, I conclude that what I have shown in this chapter isn't primarily the result of sloppy journalism. This conclusion should become even clearer in chapters 22 and 23.

10. "Is There Life on Mars?"

In this chapter, we will briefly review some of the published statements from NASA regarding their discovery of life on Mars – in one form or another. As you will see, the picture that emerges is confused and muddled. It does not represent "good value for money." If someone were to say, "NASA is hiding something," I would agree.

But the film is a saddening bore...

The primary mission objectives for Viking 1 and Viking 2 were to obtain high resolution images of the Martian surface, characterize the structure and composition of the atmosphere and search for evidence of life. Viking 1 was launched on August 20, 1975 and arrived at Mars on June 19, 1976. Viking 2 was launched on Sep 9, 1975 and arrived at Mars on 08 July 1976. The total cost of the Viking project was roughly one billion dollars.

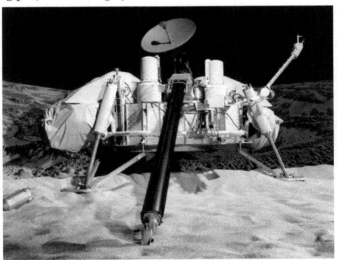

Viking Lander with extended digger arm.

Take a Look at the Law Man Beating up the Wrong Guy...

I remember back in the late 1970s, the results of the Viking Soil Sampling Experiment (called "Labelled Release" – LR) were announced – the experiment had successfully detected Microbes in the Martian sand. This seemed significant but not all that exciting. Great – there were some microscopic bacteria metabolising chemicals in the Martian soil or atmosphere. Then, NASA "unannounced" this discovery, about 1 year later, saying the chemistry behind the life-detection experiment was flawed and what had been detected was the result of some kind of inorganic oxidation effect. We were "back to square one" (or should that be 'quadrat' one). Mars was a dead world. I never even thought about this issue until about 25 years later.

But what is "Labelled Release"? On his website[231], and in a video posted on 19 March 2014, on the SPIE[232], scientist Dr Gilbert Levin explains how he invented an experiment whereby he would dope nutrients with radioactive Carbon (14). These nutrients can be fed to bacteria and then when the bacteria consume the nutrients and undergo respiration, Carbon Dioxide is given off. This carbon dioxide, however is slightly radioactive, so can be detected. Hence, if you gave some of the nutrients to "Martian bacteria" (found in a sample of Martian soil, dug out by the lander) and then your experiment detects radioactive CO_2 it means bacteria have metabolised the nutrients – and hence they must be alive. Levin explained how they were careful to make sure they repeated the experiment after heating a sample to 160°C to see if they got the same result – they did not. It seems that ever since the "life has been found!" result was announced, there have been ongoing attempts to not only cast doubt on this result, but actually debunk it. This is, in part, because there were three separate life-detection experiments built into the Viking Landers and one of them, the gas chromatograph mass spectrometer (GCMS) produced a negative result.

In a 2001 paper, Levin discusses some of the "debunking" attempts[233]:

None of the many attempts to establish the oxidant's mimicry of the LR data did so. Nonetheless, their results were deemed sufficiently similar to the LR's to validate the oxidant theory. Ironically, the only "evidence" for an oxidant in the LR sample was a re-interpretation of the very type of result that, prior to the Viking Mission, had been generally accepted as proof of an LR detection of life. Over the years, this circular reasoning was used to form a strong consensus among the scientific community in favor of the oxidant theory, even in the face of accreting contrary evidence.

In other words, over the years, Levin became more and more convinced that his experiment produced an accurate, positive result. In the video in the posting on SPIE (the International Society for Optics and Phonics)[232], he states:

Since my experiment [in 1976], NASA has refused to send another life detection experiment to Mars - or any place. Every time I have sent in an upgrade of my original experiment, as a proposal for a new spacecraft, it has been rejected. I was told directly by the head of the Mars programme, "If you send in another proposal to send an experiment to look for life on Mars, it will be immediately rejected." That's bad science. The reason they give is because if they got another equivocal answer [they said] ... "it would hurt our programme drastically, so we're not gonna take a chance." That is not science, it's politics.

As we shall see, the exact same pattern of ambivalence regarding the detection of the signs of life on Mars has been repeated several times, outside of the Viking experiments.

'Cause I wrote it ten times or more...

ESA's Mars Express probe carried an instrument called the Planetary Fourier Spectrometer (PFS) which was able to detect methane gas in the atmosphere. Even more significantly, it could detect another "biomarker" gas, Formaldehyde. Indeed, this is exactly what happened. In March 2004, Formaldehyde was detected in the Martian atmosphere. Its presence, in measurable quantities, is significant as it decomposes quickly. For example, if you were to put an amount of Formaldehyde in the atmosphere, say released from a bottle, it would decompose into other compounds, through the action of sunlight, in only a few hours[234]. This therefore suggests that something on the surface or in the atmosphere of Mars is producing formaldehyde. Formaldehyde is an organic compound – thought only to be produced as a by-product of active biological processes. Its presence, then, could indicate existing life on the Red Planet.

In an article on the BBC News Website[235], dated 25 February 2005, it was reported that an Italian scientist named Vittorio Formisano, who was working on the Mars Express probe, said "gases detected in the planet's atmosphere may indicate life exists on the Red Planet today." Formisano had told a Dutch space conference that methane and formaldehyde could signify biological activity. He was quoted as saying:

"[My observations] should not be taken as a statement that there is life on Mars today, because we need to go there, to drill the soil, take samples, and analyse them before possibly concluding that life is there."

This is a very peculiar thing for Formisano to say, because as I have already explained, this is *exactly* what was done in 1976, with the Viking lander and Levin's LR experiment etc. Why did Formisano not mention this? Why did the BBC article not mention it? The 2005 BBC article continues thus:

Professor Formisano expressed his views on the subject at the European Space Agency's Mars Express Science Conference in Noordwijk.

He said that if the methane was considered in isolation, it appeared too small a source to be biogenic in origin.

However, he argued, if the formaldehyde detected in the atmosphere was viewed as a by-product of the oxidation of methane, it would imply much more methane was being produced each year - and this could be explained by life more easily.

"If you consider only methane which is observed in the Martian atmosphere, it would be 150 tonnes a year; if you consider formaldehyde then you have 2.5 million tonnes [of methane] per year, which is much more," he said.

"And the correlation indicates the sources are in the soil, underground."

> *PFS data shows that the highest concentrations of methane overlap with the areas where water vapour and underground water-ice are also concentrated.*

These results were reported on BBC radio news in July 2005, but it seems they have not been followed up and later reports since 2005 don't seem to have referenced these results.

It's The Freakiest Show...

On 15 January 2009, a report appeared in the Belfast Telegraph[236] entitled, "NASA: Methane clouds could be evidence of life on Mars." The report was so brief, I have included all of it below.

> *We could be about to discover that we really are not alone.*
>
> *NASA's holding a news conference amid claims craft orbiting Mars have [detected] clouds of methane gas.*
>
> *The evidence could point to life of some form on the Red planet, possibly organisms existing below the surface.*
>
> *Some form of volcanic activity or unknown chemical reaction could also be the cause.*
>
> *But even if the methane was produced biologically, it could be leaking from some ancient underground source from the far distant past.*

This seems to contain less detail than the BBC report from 2005 – almost 4 years earlier...

A now-difficult-to-trace article on NASA's Website also dated 15 Jan 2009 has a headline[237] "The Red Planet is Not a Dead Planet." It confirms the same details about the likely biological source of the Methane. It includes the graphic shown below, and includes the following explanation:

> *If microscopic Martian life is producing the methane, it likely resides far below the surface, where it's still warm enough for liquid water to exist.*

Again, the article makes no reference to the Viking LR experiment, nor does it refer to earlier Mars Express methane and formaldehyde data.

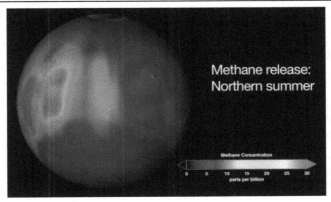

Above: Methane plumes found in Mars' atmosphere during the northern summer season.[237]
Credit: Trent Schindler/NASA [animation]

Oh man, wonder if he'll ever know!

Another three years pass and it's 4 December 2012. An article in Wired Magazine Reports "Curiosity Rover detects water and organic compounds on Mars."[238] By this time, the next fancy billion-dollar Mars lander/rover had been built and despatched to Mars, it could now be revealed…

> *"After two weeks of speculation following Caltech geologist John Grotzinger's announcement that the Curiosity Rover's Martian soil analysis data would be " one for the history books" the results of the soil scoop are finally in.*
>
> *According to Nasa the rover's onboard laboratory found water, sulphur and chlorine-containing substances in the sample taken from the Rocknest site in Mars' Gale Crater. The chlorine-containing substances include the chlorine and oxygen compound perchlorate which had previously been found in arctic Martian soil by the Phoenix Lander.*
>
> *The rover also detected its first organic compounds in the form of chlorinated methane created during the analysis process within the Sample Analysis at Mars (SAM) tool. However, while the chlorine is definitely from Mars, the carbon might be of more Earthly origin, carried to the planet by the rover."*

And then, less than 10 months later, on 20 Sept 2013, another article appeared in the UK Daily Mail, entitled "NASA says there is NO life on Mars,"[239] with the following key points:

- *NASA says there is NO life on Mars: Curiosity rover hasn't discovered any clues that the atmosphere supports living things*

- *Curiosity rover has found no sign of methane, which is produced by life*

- *Robot has spent a year on Red Planet scanning its surface and atmosphere*

- *'If you had life somewhere on Mars, you might see some', NASA says*

The article stated:

> After a year roaming the surface of Mars, Nasa has failed to find any evidence that its atmosphere is supporting life, it was revealed today....
>
> Mars today is a hostile place - extremely dry and constantly bombarded by radiation. Billions of years ago, the planet boasted a thicker atmosphere and possible lakes. Scientists generally agree that nothing can exist on the Martian surface at present since it's too toxic. If there are living things on Mars, scientists theorize they're likely underground.
>
> Just because Curiosity didn't detect methane near its landing site doesn't mean the gas is not present elsewhere on the planet, said Bill Nye, chief executive of the Planetary Society, a space advocacy group.

Again, there is no mention of earlier reports, referenced above. Let's go round, one more time, just for fun...

He's in the best-selling show...

An article on Wired Magazine's website, dated 17 Dec 2014 declares: "Curiosity detects first burps of potential life on Mars."[240] The article notes:

> Large quantities of methane have been detected by Nasa's Mars Curiosity rover, suggesting there might be potential for life on the Red Planet. Nasa said that tenfold spikes in the levels of methane had also been detected, hinting further at microbial organisms.

It repeats:

> While the discovery of methane could be biological, it could also be non-biological

NASA's own website also reported "Curiosity Detects Methane Spike on Mars"[241]

> NASA's Mars Curiosity rover has measured a tenfold spike in methane, an organic chemical, in the atmosphere around it and detected other organic molecules in a rock-powder sample collected by the robotic laboratory's drill.

Yet again, there is no reference to earlier results from 1976, 2004, 2009 or 2012. Over 10 years *after* the Formisano statements, NASA still haven't figured this out, according to them. Shocking.

Conclusions – "She could spit in the eyes of fools..."

This chapter documents a very similar pattern to what we saw in the previous one – media reporting of evidence of life Mars discovered over the last 40 years. The conclusion I draw is the same NASA, and academia are covering up the discovery which was usually the Mars missions stated goal.

11. "Look at those Cavemen go"

Bones of Contention

Since the Spirit and Opportunity rovers started operating on Mars in 2004, a few independent scientists and engineers have closely scrutinised the images they have returned. Just as we have seen in the preceding chapters of this book, NASA have taken little or no interest in their findings.

We will start off with an image from the Curiosity rover that was posted on the UFO Blogger website[242]. (We should note that, because it was posted on there, it is of no further scientific value to NASA or anyone else, if we use the logic that we will encounter in chapter 22.)

On the UFO Blogger Site[242], this was likened to a Dorudon Atrox Fossil found in Wadi El Hitan (Whale Valley), Al Fayoum, Egypt[243]. The Curiosity image (above) was taken on 25 Nov 2012[244].

Stromatolites?

In his free eBook, called "The Living Rocks Of Mars," Lyall Winston Small[245] has made an argument that some rover images show stromatolites [246]– a type of sedimentary rock whose structure and appearance is greatly influenced by either the action of bacteria or by the action of organisms like lichen. (The book is quite well put together, though the referencing seems a little haphazard and many images seem too hard to find.) With one type of stromatolite, the rock can have a "layered" or "laminated" appearance. Thrombolite - another type of stromatolite - has a clotted structure or a pattern is formed by iron-eating microbes. The image comparisons in his book are quite compelling.

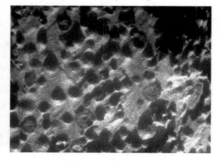

Cambrian Era Stromatolite rock[247]

Mars Rover image
1M400198734EFFBVM5P2956M2M2[248]

Destroying the Evidence?

In the Hoagland/Bara "Dark Mission" book, the image below is mentioned – does this show another type of fossil? The sample was destroyed by the Rover's drill.

Far Left:
Opportunity
Rover Image of
something that
looks like a
Crinoid Fossil[249]

Left – a Crinoid
fossil on Earth

Martian Fossils and "A Modern-Day Einstein"

Sir Charles Shults III is a US Scientist and researcher who worked at Martin Marietta Aerospace for 10 years, on software for Cruise Missiles.[250] He has spoken at length, several times, on the US Coast to Coast talk show[250] about his analysis of a number of photos from the Mars Rovers. A selection of these images, some of which we will discuss below, can be found on his website - http://www.shultslaboratories.com/

Sir Charles Shults III appeared on his local TV station, WLTX, on 29 April 2011[251].

As well as evidence of liquid water on the surface (see chapter 9), Shults seems to have found what look like small, fossilised sea creatures in some rover images. Shults was interviewed by his local TV news channel in April 2011. In that interview, he stated he was certain he had found evidence of liquid water on Mars on 15 February 2004. Having obtained more than 200,000 Mars images from NASA under a Freedom of Information request, he spent thousands of hours looking for small details. He then posted some of his results on the internet. He was interviewed about this on 21 August 2005[252] on "Coast to Coast".

He has not been able to get anyone at NASA to publicly engage in the debate about his findings, and he states he was once "bounced" from a scheduled conference appearance, where he would have presented these findings. Hence, in 2008, he produced "The Fossil Hunters Guide to Mars[253]" as a collection of PDF files, on a CD.

One of the clearest images seems to be of a spiral sea shell fossil:

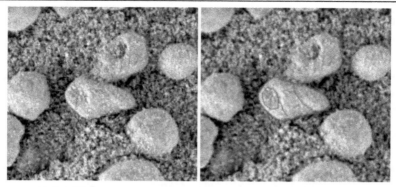

More pareidolia? [254]These objects can be found in an Opportunity Image from Sol 505[255].

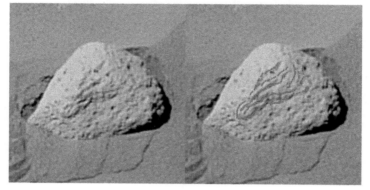

Shults suggests this is a Crinoid fossil - identical to those found on Earth[256]. This can be found on a Spirit Rover image from Sol 343[257].

Shults suggests this is a Sea Urchin fossil - a small spherule showing the star marking on its surface, identical to many sand dollar and sea urchin species on Earth[258]. This can be found on an Opportunity Rover image from Sol 507[259].

Shults suggests this is a Copepod fossil[260] – a type of crustacean[261]. This can be found on an Opportunity Rover image from Sol 239[262].

Shults suggests this is the fossil of a broken trunk/stem of a plant[263]. This can be found on a Spirit Rover image from Sol 007[264].

On his website, Shults echoes the sentiments of Dr Brian O'Leary, which we covered in chapter 4:

> *The most interesting thing about these findings is this - even though these objects have been seen for years, and even though three different rovers have literally driven right up to them or even over them, in all cases NASA has chosen to drive away without addressing the most basic question: what are we seeing?*
>
> *In this mission of exploration and learning, we are finding hundreds of anomalies and mysteries, and many of them are so obvious and striking that even a child would ask what they are seeing. Yet, even with some of the best equipment on site and some of the best scientific minds looking at these findings, not one has asked what they have found.*

In his e-book "The Fossil Hunters Guide," Shults expands on the above findings considerably and he also goes to great lengths to show much more evidence from the Spirit and Opportunity Mars Rovers of features that can only have been created by liquid water (from small "geysers," he suggests).

Bolder than Bolden?

In the April 2011 news report, referenced above, the interviewer also spoke to former Astronaut and then administrator of NASA, Charles Bolden to ask him about the question of life on Mars. The news report said:

> Charlie Bolden won't speculate on the potential for water or fossils on Mars but he's hoping the next rover mission, Curiosity, will answer those questions once and for all. It takes off this November. The first data is expected by the fall of next year.

Bolden then went on to speak about the (then) upcoming Curiosity/Mars Science Laboratory mission and said to the reporter:

> We expect to make discoveries with Curiosity. I don't want to, you know, set great expectations. What we find will be different from what we found before. It will look at the soil of Mars. It's looking for any signs of microbes or microbial life there. It's absolutely incredible what its gonna do...

The reporter quizzed Bolden about Shults findings and Bolden replied:

> I don't know, you know? I wish I could say "yes, I do," but I don't know. I really don't.

(We have already covered, in chapter 10 the flip-flopping that has been done with statements about the Curiosity rover, which were almost exactly the same as the flip-flopping that was done with the Viking data, more than three decades earlier.)

In the April 2011 news report and interview, Shults makes a good point, which is supported, I would say, by the evidence discussed in chapter 10.

> Nothing we know of would have sterilized the entire planet - funguses spores and bacteria could very easily still be in the soil today.

We will examine further evidence relating to this statement in later chapters.

False Colour

Starting on page 75 of his "fossil hunters guide" Charles Shults has also documented evidence of image tampering, which we will discuss in chapter 13.

A Further Paper on Past (and Present) Life on Mars

A paper by Rhawn Gabriel Joseph, posted on an independent website called Cosmology.com[265], draws similar conclusions about the presence of fossils and other rock forms on Mars. It shows several other Rover photographs that are not shown in this chapter and also considers some of the evidence we will cover in the next chapter.

Conclusions

As with the investigation of Cydonia and the Face on Mars, covered in chapter 4, we have seen in this chapter how some excellent and comprehensive independent research, using NASA's own data, has revealed evidence of Life on Mars (if the rovers are on Mars). NASA has ignored all of this research and you will not find anything of substance written about this evidence on any of NASA's websites. We will see the exact same situation with evidence we will examine in the next chapter.

12. There is Life on Mars

We have already looked at fossil evidence, water evidence and gas evidence. Is there any other evidence that life exists on Mars? The answer, is "yes, absolutely..."

The inventor of the Communications Satellite

Another person that was paying attention to the MGS images that started coming down at the turn of the millennium was someone who was associated with the turn of the millennium itself. Arthur C. Clarke[266], author of "2001: A Space Odyssey" and many a science fiction novel and essay, saw the image below.

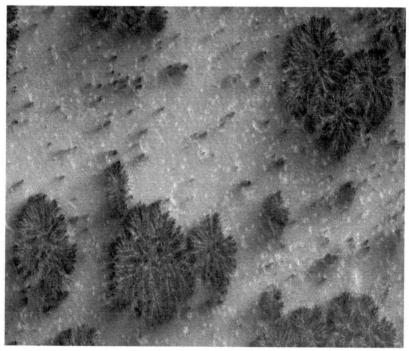

A portion of MGS Image M0804688[267] (rotated to fit on the page better) – Taken on 19 Oct 1999[268]

Clarke gave an "email interview" for the "Space.com" website (which is now unavailable and does not come up when searching the site), that was posted on Clarke's own Website on 7 June 2001[269] (this is also only retrievable from the Internet Archive/Wayback Machine). It was reported that he said:

"I'm quite serious when I say 'have a really good look at these new Mars images.' Something is actually moving and changing with the seasons that suggests, at least, vegetation"

Clarke repeated several times that he was serious about his observations, pointing out that he sees something akin to Banyan trees in some MGS photos.

(The story was, a few months later, covered briefly on the "Popular Science" website.[270]) The field of view in the image is quite large – the portion shown above is about 2 miles across. So, if these were trees, they would be quite large in their extent.

Clarke, who under other circumstances, would be taken very seriously, was largely ignored. James Rice, a planetary geologist at Arizona State University said:

Most of the images in question that I looked over are of dunes in the polar regions. They are beginning to defrost as winter draws to a close. The sand composing the dunes is dark and frost is bright, thereby causing the spotted pattern as the dune defrosts. This is not vegetation but rather the natural defrosting of dunes and sand sheets.

A similar image was documented on the MSSS website on 10 August 1999[271].

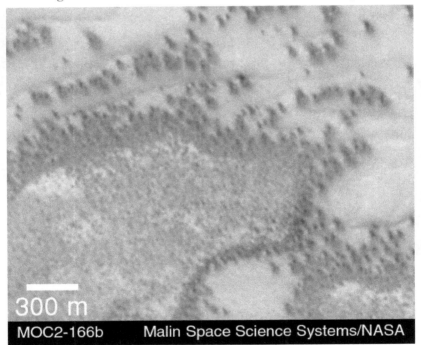

MOC2-166b   Malin Space Science Systems/NASA

Are these bushes?[272]

A confident explanation is given for what we are seeing:

Because the Martian air pressure is very low--100 times lower than at Sea Level on Earth--ice on Mars does not melt and become liquid when it warms up. Instead, ice sublimes--that is, it changes directly from solid to gas, just as "dry ice" does on Earth. As polar dunes emerge from the months-long winter night, and first become exposed to sunlight, the bright winter frost and snow begins to

> *sublime. This process is not uniform everywhere on a dune, but begins in small spots and then over several months it spreads until the entire dune is spotted like a leopard.*

Yet this was long before the presence of surface water was acknowledged. We discussed in chapter 9 how it wasn't until September 2015 that we heard someone at NASA state clearly they had found evidence of liquid surface water. These explanations therefore seem to be rather self-serving. It continues thus:

> *The early stages of the defrosting process--as in the picture shown here--give the impression that something is "growing" on the dunes. The sand underneath the frost is dark, just like basalt beach sand in Hawaii. Once it is exposed to sunlight, the dark sand probably absorbs sunlight and helps speed the defrosting of each sand dune.*

Several similar photos exist, such as MGS image MOC2-286, taken on 12 June 2001[273], a portion of which is shown below:

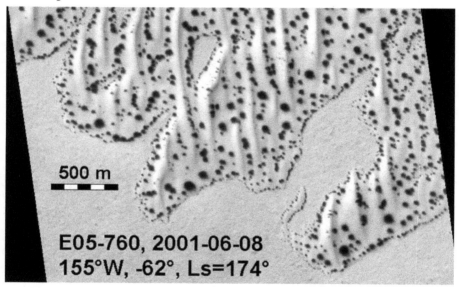

Of course, we must be seeing more defrosting dunes here[274], nothing like cultures of bacteria *growing* on agar jelly in a petri dish…

Rings on Mars – an Independent Analysis

Another independent researcher – Alan D Moen has also investigated images like these[275]. In 2004, the Lunar and Planetary institute [276] published a paper he submitted which summarised his early findings[277]. Later, in June 2006, he posted a remarkable YouTube video[278]. In this video he studies "ring features" on MGS images, which appear to have changed shape over time. In his video, about the changes in appearance of these rings, he states.

> *NASA has proposed explanations - such as exploding gas and springtime thawing of frozen CO_2, but these theories usually come with the qualifiers*

"speculation," "thought to be," "much remains unknown" and "not well understood. The existing theories raise more questions than answers there is, however, another possibility - an explanation that's known to match ring shape to date/time, latitude and surrounding surface features, at multiple locations."

The video illustrates a scientific approach he has taken in that he has found MGS images of the same areas, but taken at different times – when the sun angle is different. In each of the pairs of images, Moen has noted how the shapes of rings has changed. Then, using trigonometry based on (a) the measured length of the shadows and (b) the known angle of the sun (which is given with the image data for the MGS images), he has calculated the lengths of the shadows. showing features similar to these – and other features such as strange rings.

Moen examines image SP2-53807[279], as shown below:

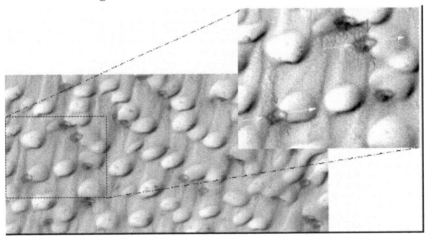

Alan D Moen's annotation[275] of MGS image SP2-5307[280] (84.73°N, 0.89°W)

He notes:

[The image] contains hundreds of rings in a field of small dunes. It was observed, in this image, that the orientation of ring elongation matched the direction of sunlight. Ring, after ring, after ring - precisely aligning with the Sun. In addition to the alignment with sunlight direction, ring distortion or directional change was noted, in cases where the ring is transitioned from flat ground to slope dune surfaces. These two qualities imply that the dark rings are <u>shadows</u>, cast by features on the surface of the planet.

Moen built a physical and computer model to try and reproduce the "behaviour" of these rings. This is all illustrated and explained very clearly in his YouTube video.

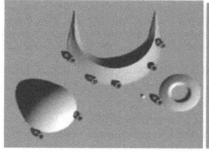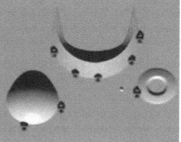

Moen's computer modelling of the ring appearance with changing shadows.

Physical Model shown in Moen's 2004 paper.

Resulting "Ground View" of objects creating shadows.

On his webpage[275] and at the end of the video, Moen contends that the conditions could be suitable for trees to grow, exactly where these ring features appear:

A layered construction, similar to a spruce tree, creates both bright reflective rings and dark shadow rings that closely match the various Mars Global Surveyor MOC images. All the elements needed for life are available at the northern polar locations - high concentrations of water in the form of ice, carbon dioxide and sunlight. The recent discovery by the European spacecraft Mars Express of methane in the Martian atmosphere lends more support to the possibility of present-day life on Mars.

How refreshing – Moen refers to other Mars data to back up his conclusions! That is, we covered the methane evidence in chapter 10. For me, Moen proved his case. The simplest explanation for the effects he observed is that trees are growing in that area.

In a 2005 posting on the "SETI At Home" forum[281], Moen gives a timeline of his research. For the Jan-Feb 2005 period, he states:

Emailed over 500 scientists and students at various universities around the county and world. Received some positive response and managed to get around 150 people to view the movie. But the majority feedback was: life does NOT exist on Mars, so this CAN'T be what you say. That sounds more like religious dogma to me than science. No one offered to prove me right or wrong by doing a peer review of my work. Wasn't anyone curious?!

After this he writes:

I decided there was a huge scientific bias that life did NOT exist on Mars, and that scientists would not give an unbiased analysis of the NASA images. So I decided to send MGS MOC dune/spot images to USGS remote sensing scientists without telling them they were Mars images. I gave them the scale/size and asked them if the spots were dark dust, tree shadows, or something else. Four scientists responded that the spots in the Mars images were shadows and or bushes/shrubs.

This mirrors closely what I experienced with getting scientists to comment on the "dome" image covered in chapter 7, although Moen did a far, far better job with his research and he was more persistent than I was. On his website[282], Alan Moen posts several of the positive comments he received, such as this one from a member of the Geographic Science Team, EROS Data Center:

"...it does appear that the dark spots are some form of vegetation, since they are on the lee side of the dunes, they are in a cooler and perhaps more moist environment - it appears that there are few on the predominately sunlit side of the dunes. The bottom image looks a lot like some that we see in Iraq and Afghanistan - the dark spots here also appear to be trees or shrubs - very sparse, most likely because of the low levels of precipitation and organic soils."

Since Moen produced this excellent piece of research, we have seen new higher resolution images from MRO. Some them also show the same sorts of things that Moen observed. By now, it shouldn't be difficult for you to guess how this has been reported.

A "Tantalizing Tree Illusion"

On 13 January 2010, Space.com posted a short article entitled "Strange Mars Photo Includes Tantalizing 'Tree' Illusion[283]." It included the high-resolution image shown below:

The caption for the image reads: CREDIT: NASA/JPL/University of Arizona. "This new image of Mars taken by NASA's Mars Reconnaissance Orbiter shows an optical illusion. What appear to be trees rising from the Martian surface are actually dark streaks of collapsed material running down sand dunes due to carbon dioxide frost evaporation. The image was released in Jan. 2010." It was taken at Latitude 83.506°N, Longitude 118.588°E

In this particular article, no image number or data is given, and there is no analysis. As usual, most people will accept the "cold, dead, lifeless explanation," because unnamed NASA scientists tell us that is the only explanation that is sensible.

This image was featured in the mainstream press such as the UK Daily Telegraph[284], UK Daily Express[285] and the Wall Street Journal[286]. After some research, I found a link on the APOD website[287] to the HiRISE data page for the image – number PSP_007962_2635.[288]

How many people that saw this image actually tried to find the original image data and then do research on it, as Moen did, years earlier with MGS images?

Mysterious Motile Microbes

In researching the topics covered in this book, in 2010, I came across another amateur researcher - Ron Bennett. Like Alan Moen, he had diligently studied NASA images and come up with some startling conclusions. In 2009, he posted a YouTube video about his findings[289]. Bennett was studying images from the Phoenix Lander[19]. This lesser-known Mars mission touched down at 68.15°N,

125.9°W, on 25 May 2008 and ran for almost 6 months, with communications being lost on 02 November 2008. From the Phoenix mission's description, we read:

> *The Phoenix Mars Lander is designed to study the surface and near-surface environment of a landing site in the high northern area of Mars. The primary science objectives for Phoenix are to: determine polar climate and weather, interaction with the surface, and composition of the lower atmosphere around 70 degrees north for at least 90 sols ... characterize the geomorphology and active processes shaping the northern plains and the physical properties of the near-surface regolith focusing on the **role of water**... as well as the adsorbed gases and organic content of the regolith ... determine the past and present biological potential of the surface and subsurface environments.*

"Regolith" is basically another word for "soil."

The Phoenix Lander – Observant people will notice the "digger arm" which is very reminiscent of the Viking Probe.

Ron Bennett (left) studied thousands of phoenix images. He created colour images from the grey-scale ones posted on NASA's website[290]. He did this to bring out much more information in the images.

He created some time lapse sequences which are, to me, very revealing. Below, I have captured some stills from his video, which has had a small number of views, considering it has been online for over 8 years (at time of writing).

In the images below, consider how the positions of the "blobs" indicated change in each subsequent image (it is better if you go online and watch Bennett's video).

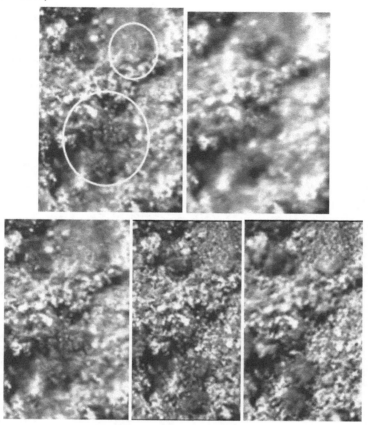

Sol 123 – Images are 47 minutes apart

Bennett notes:

> *Images from Phoenix's Optical Microscope show nearly 1,000 separate soil particles, down to sizes smaller than one-tenth the diameter of a human hair. At least four distinct minerals are seen.*

He suggests the unknown "particles" which move in the sequence of Sol 123 images shown above could be tardigrades[291]. From Britannica.com, we read:

> *The most remarkable feature of the tardigrades is their ability to withstand extremely low temperatures and desiccation (extreme drying)*

A 3D electron microscope image of a Tardigrade[292] – approx. 50μm x 60μm

In another sequence, we also see movement and even "flexing" of "something"

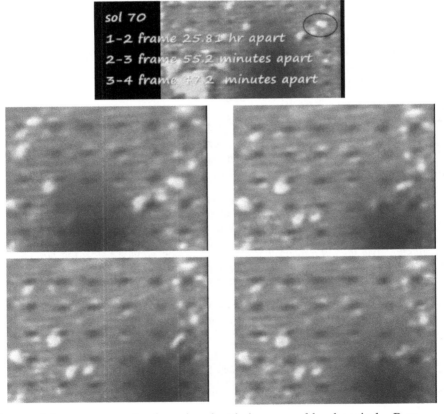

Some might say we are simply seeing dust being moved by the wind – Ron Bennett points out:

A dust devil would have that energy, but these images are shot minutes apart, so if a dust devil moved over the rover, it would be gone in a flash and wouldn't hang around for several minutes to several hours during the time the first image was shot and the last image was shot.

That is, the movements seen here take place over minutes or hours, not 1 or 2 seconds. I really recommend you attempt to view Ron's videos online, to see the effects he has documented.

Once again, I cannot find any relevant comments from NASA – and I don't think anyone from NASA has publicly commented or attempted to reproduce Ron's findings – which are simply from NASA's *own* images – as with almost everything else in this book!

In a later video that Ron Posted on 20 October 2012[293], he usefully pointed out this:

In 2008, the Phoenix lander science team claimed that all the necessary nutrients to sustain life were found at the Phoenix lander site. In fact, they claim that this is a type of soil you probably have in your backyard - you might be able to grow asparagus pretty well…

Bennett also (in his softly-spoken manner) discussed the issue of Perchlorates in the Martian soil that Phoenix sampled and concluded that based on what the Phoenix Science team had found in performing the lander's experiments, the objects he highlighted could be living organisms.

Growing Asparagus on Mars

On 26 June 2008, Scientific American carried an article entitled, "Pay Dirt: Martian Soil Fit for Earthly Life.[294]" The subtitle read "Phoenix finds alkaline soil with plenty of minerals."

Mission scientists say the soil has a pH between 8 and 9, which places it somewhere around seawater or baking soda in alkalinity. It also contains the minerals magnesium, sodium, potassium and chloride. Further analysis is expected to reveal whether it contains other chemicals such as nitrogen and sulfates. The finding implies that life could indeed survive below the surface, where it would be protected from harmful ultraviolet rays and harsh oxidants that might accumulate on the top layer of soil.

As I have noted elsewhere in this book, this article makes no reference to Viking, Mars Express or any of the other images or data which indicate that life was found on Mars 40 years ago.

It's Just another Image Processing Effect…

In 2004, ESA published a new image of the Gusev Crater taken by Mars Express. Unless you are seeing this book in colour, you will only see a grey streak in the image below. In the full image, the patch is **green**.

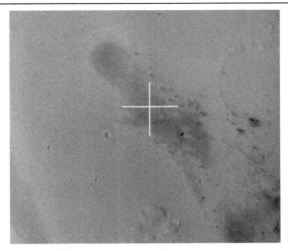

*The centre of the Gusev crater with the landing site of the NASA Spirit rover marked with a cross. The image was taken by the HRSC instrument in colour and 3D on 16 January 2004 from a height of 320 km. Gusev is a large crater about 160 km in diameter. Scientists believe that the crater was covered by standing water, maybe in the form of a lake, early in the history of Mars. This is a section of a larger picture, accessible as a high resolution download, which covers 60 m from top to bottom. North is at the top. **(Note the green colouring is an effect of image processing)***

Already, then, ESA are playing down the significance of the green colouration. According to a page on Richard Hoagland's "Enterprise Mission" site, Michael McKay, Flight Operations Director of the European Space Agency, on seeing this picture said[295]:

> "... like the green in the Gusev crater picture ... it certainly gives rise to the speculation that there could be algae [there] .... It certainly gives much more weight to such speculation, particularly since here on the Earth's glaciers and [in] the Alps and [at] the North Pole, you can see algae in the ice itself which turns rather a pink color or greeny-grey color. Just tying that observation on the Earth together with things we are starting to see on Mars, certainly adds a bit more weight and people will seriously be thinking about these questions and trying to put some definite answers to them ...."

The interview with McKay is behind a "Paywall" on Linda Moulten Howe's "earthfiles[296]" website, however.

Hoagland also notes that another version of this image appeared on the German Space agency's website about the same time (retrieved from the internet archive)[297]. Again, you will likely only see how different this image is if you are viewing this in colour.

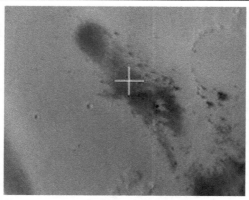

Mars Express Gusev Crater Image found on German Space Agency Website showing different colouration.[298]

One is also given to wonder what is being shown in another Mars Express image, posted on 23 Jan 2004. (Again, you need to see this in colour.)

This picture was taken by the High-Resolution Stereo Camera (HRSC) onboard ESA's Mars Express orbiter, in colour and 3D, in orbit 18 on 15 January 2004 from a height of 273 km. The location is east of the Hellas basin at 41° South and 101° East. The area is 100 km across, with a resolution of 12 m per pixel, and shows a channel (Reull Vallis) once formed by flowing water. The landscape is seen in a vertical view, North is at the top.

This page does *not* state the colouration is due to image processing…

If you were to download a higher resolution version of this image, you can zoom in and, for example, look at this:

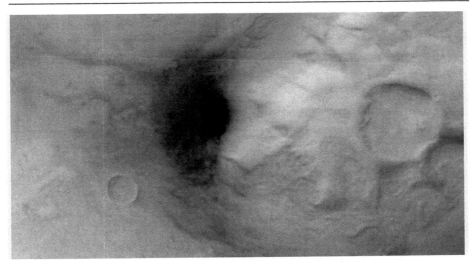

Enlarged section of Reull Vallis Mars Express image.[299]

But, of course, we aren't seeing evidence of green vegetation, are we? It's just image-processing.

An article by Richard Hoagland[300], which we will reference again in chapter 13, states that chlorophyll *was* discovered by the Pathfinder/Sojourner lander in 1997 – and this is referred to in an article on BBC News dated 05 April 2002, "Life on Mars hopes raised."[301] This article states:

> Scientists have found "intriguing" new evidence that may indicate there is life on Mars.
>
> An analysis of data obtained by the Pathfinder mission to the Red Planet in 1997 suggests there could be chlorophyll - the molecule used by plants and other organisms on Earth to extract energy from sunlight - in the soil close to the landing site. Researchers stress their work is in a very preliminary state and they are far from making definite claims.

Conclusions

In this chapter, then we have seen images of

- What could be trees.
- What could be bushes.
- What could be small organisms moving around.
- What could be algae or something similar.

In addition, we have noted how the soils in the northern regions are, according to NASA scientists, suitable for growing certain vegetables.

My conclusion is… there is life on Mars, folks!!

13. "Hey, You with The Pretty Face…"

> **In this chapter, you will need to view the images in colour to understand what you are seeing. Hence, if you are reading the paper version of this book, please go online to https://tinyurl.com/sitsbook to download a PDF version of this book.**

If one starts to really study the alternative information regarding the images from the Mars landers and rovers, it soon becomes clear that something is amiss. Someone, again, is either not telling the truth or is being wilfully ignorant or negligent regarding the enormously expensive science projects they have been involved with.

In this book, up to now, we have primarily dealt with monochrome/black and white images. In public presentations I have given about this topic, I have often asked why the vast quantity of Mars images taken after, say, the turn of the millennium, were not released in colour. My rationale for saying this comes from the knowledge that the first camera phone was sold in year 2000[302]. By 2008, smartphones with cameras had become quite common, and were capable of taking quite good photographs. Hence, one would think that with "billion-dollar budgets," NASA could afford high-tech high resolution and high-performance colour digital cameras that would work on the surface of Mars. They should have had such equipment built *years* before such technology became part of the paraphernalia of our daily lives.

That said, I can appreciate that for certain scientific reasons, such as getting the best range of imaging options (e.g. looking into the infra-red or ultra-violet ranges), a sensor with black and white imaging will work best. Also, having changeable filters, placed in front of a black and white image sensor, can give more control over the final image.

It is also true that, particularly since the 2012 curiosity rover, more Mars colour images have been made available.

However, as with much of the other subject matter in this book, the colour images from Mars have been the subject of independent, careful and informed scrutiny – not just for the anomalies they show, but for the *colours* seen. In researching anomalous Mars images, I came across a number of separate analyses of NASA image data which seem to strongly suggest that the colour of the Martian Sky could be quite different from the murky red colour which appears in the vast majority of the photos we have seen from the surface of the planet.

Indeed, the colour of the sky, as seen from the surface of a planet, is not dictated by the colour of the ground. That is, just because Mars has a red surface should not mean that the sky would appear red – when seen from the surface. The blue colour of the sky we see here on earth is caused by Rayleigh scattering – whereby gases in the atmosphere scatter most of the blue light from the sun's rays. We only normally see a red sky at sunset because we are seeing the sun through *more* of the earth's atmosphere. Hence, more of the blue light is scattered, so we are left with much more red light.[303] The same isn't necessarily true on Mars – because the atmosphere is much thinner than here on earth.

"Mr Blue Sky…"

Even Carl Sagan seemed to question the initial findings about the colour of the sky on Mars. At the Viking Lander press conference on 21 July 1976, he stated:[304]

> *Dr James Pollock of the lander imaging team has now looked at individual brightness numbers so-called DN values and concludes that despite the impression on these images, the sky is not blue. The typical, typical earth chauvinist response. [Laughter] According to Dr Pollock the sky is, in fact, pink. Which is an okay color. The sky is red, but it is not as red as the surface and the question of what makes… the reason it looks blue here - it must then be due to someone making a slightly wrong relative weighting of the three colors which go into making up this picture, and it was done before the appropriate color calibration chart data had been incorporated into this, so I hope we will see in the next day or two it corrected….*

Was he joking? Even if he was, it sounds like he was looking at images that had more of a blue sky than a pink or red sky…

Richard Hoagland's Enterprise Mission site was one of the first I came across which seemed to contain relevant evidence that the colour of the sky shown in various colour images from Mars was not correct. In an article called "Revealing the True Colors of NASA," Hoagland claims that when Viking landed on Mars, the initial pictures showing a Grey/Bluish sky (left) were deliberately changed, in the first 2 hours, to show the reddish colour (right)[305]. Barry E. DiGregorio and Dr Gilbert V. Levin corroborate this story in a book called "Mars: The Living Planet.[306]"

Changing the colours of the sky on Mars? The right-hand image can be found on NASA's website[307].

A follow up article that Hoagland authored, contains more detail about the way "colour should work" in the Mars images[300]. Hoagland also references later missions. For example, in relation to what we already covered in chapter 12 about the possibility of algae growing on Mars, he suggests that certain filters used on the Mars Odyssey's THEMIS instrument would not be sensitive to the light reflected by chlorophyll – the compound in green vegetation which is responsible for photosynthesis.

"Please Tell Us Why…"

After several years of being aware of "problems" with the Mars image colour balance, I came across a BBC Horizon Documentary about the Viking mission, first aired in 1977[308], which seemed to confirm what Hoagland's page said. Here is a still from the documentary, along with a transcript of some of the commentary.

*On the second day the first pictures in colour. Mars was indeed the red planet - or at any rate, blue-red. But the colours were false - **it was noticed that one of the cables on the spacecraft wasn't orange enough.** The colour was corrected and Mars turned even redder. The reason for their mistake?*

*The sky - **everyone had assumed it would be blue, like on earth**, but on Mars scattered dust turns the sky pink and makes sunsets purple.*

This is an utterly false explanation – because as far as I know, the colour of cables is *not* used to calibrate images on the Viking Lander. The lander had a "camera test target" – as we can see in the schematic, and photo below:

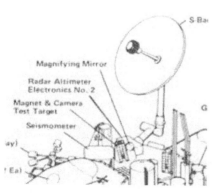

Viking Lander Partial Schematic[309]

Photo of Lander's top – showing the Camera Test Target (Viking image 12e018)

"You Had to Hide Away for So Long…"

Holger Isenberg has taken some of the original Viking Lander image data[310] and re-processed it, based on the "filter response" data he found. To explain this a little, a given Viking image (and most if not all other colour images from Mars) are generated from 3 separate grey-scale images. One image is taken through a red filter, one through a green filter and one through a blue filter. The filters have a certain "profile," which is then used when the set of R-G-B images are combined to produce a true-colour image. As we will read below, we are told that doing this from images on Mars is "difficult" because the lighting is different there – due to the atmosphere being different – and other factors. Colour calibration is, apparently not straightforward.

However, Isenberg went ahead and did his own image processing[311]. (This processing should be possible in software packages like Adobe photoshop, GIMP and Paint Shop Pro). We can see an enhanced image he produced:

From Isenberg's analysis: Viking 1, Nr. 12b069, 29 August 1976, 12.65 local Mars time. This picture was created with color-correction derived from the filter response data. [312]

Holger Isenberg also reprocessed an image we saw in chapter 9:

Left:

Viking 2 Image, 21i093, 18 May 1979, 14.24

Dr Gilbert Levin – who developed the Labelled Release life detection experiment mentioned in chapter 10 – has also written quite extensively about Mars image colour calibration. Dr Levin had a paper published in SPIE Proceedings 5163, 19, August 2003, with the title "Solving the color calibration problem of Martian lander images"[313].

He explains why there are seemingly problems with colour calibration and suggests.

> *A more technically advanced and more expensive method of determining the absolute coloration of the Martian surface would be to use a hyper-spectral imager instead of a conventional red, green, and blue color camera. A hyper-spectral imager provides the complete spectrum of each pixel leaving no question as to that pixel's color content.*

He continues:

> *Using these advanced technologies should close the broken links in the chain of calibration. Images could then be produced that would be of greater value to geologists, chemists, and biologists studying Mars. These calibrations could also be used to understand the reflectivity of the Martian scene and to produce another set of images that show how the scene would look if illuminated by light on Earth. The full potential of Martian lander imaging could then be realized.*

Levin seems to be saying that the colour images aren't very useful. However, he then says images could be recalibrated using his method based on Earth conditions…

> *In the meantime, the existing images of the Martian surface may be recalibrated to Earth conditions, as shown in Figure 9, to provide what is very likely a closer approach to reality than presently available.*

This sounds somewhat similar to what Holger Isenberg has done. In a later paper, published in the SPIE Proceedings 5555, 29, August 2004[314], Levin covers similar ground and seems to show images similar to Isenberg's.

"Look around see what you do…"

Levin has also related the colour in Viking images to the presence of "biology." In 1978, Dr Levin and his colleague Patricia Ann Straat had a paper published in the Journal of Theoretical Biology called "Color and Feature Changes at Mars Viking Lander Site (1978)"[315]

This paper looked at several Viking Lander images taken at different times – months apart. The abstract of this paper, reproduced below, makes for interesting reading – particularly in view of what we have covered in this chapter and chapter 10.

> *Analysis of three component color pictures taken by the Viking lander camera on Mars has established color differences for the background material, the rocks and spots on the rocks. Changes in the location of greenish rock patches and ground patterns have been observed over time. A combination of wind movement of dust and dirt dropped by sampler arm operations could have*

> produced the slight changes in pattern and position. However, the observed
> patches, patterns and changes could also be attributable to biological activity.
> Analysis of six component color data on the same scene confirms the
> observations including the greenish color of the rock patches.

Sounds like more chlorophyll to me...

"Where Did We Go Wrong?"

It seems problems with Mars image colour calibration have been known about for many years (Isenberg's analysis, for example, was completed in year 2000 and Levin's earlier one was completed in 2003). Both these dates, then were *before* the "twin rover" missions to Mars in 2004.

However, we saw the same pattern of deliberately incorrect colouration – as shown here, in an illustrative image created by Goro Adachi:

Red? Blue? Why even bother to calibrate the colour?

Goro Adachi details other problems with Mars image colouration on his web page "Hidden Colors of Mars..."[316]

Of course, we as lay people do not understand the science of colour and optics, do we? So, there is (literally) nothing to see here. Back in 2004, when I became aware of this problem, I found a response to the question about red/blue Mars image colouration posted on the "Above Top Secret forum.[317]"

Kano

posted on Jan, 18 2004 @ 09:34 AM   link

This article is a brief summarised explanation of how the PanCam on the Mars Spirit
Rover operates, in relation to the strange appearance of the calibration sundial in some
pictures. The question was first raised by ATS member AArchAngel, and has been
discussed at length in this AboveTopSecret forum thread and ATSNN story:
thread

Mars Spirit Rover Picture analysis.

In this thread I will attempt to summarise my posts to the larger thread.

What are you talking about?

Ok, the initial alarm was raised after it was noticed that the color-calibration sundial
mounted on the rover, looked quite markedly different in the Mars-Panorama shots
compared to its regular appearance.

Immediately wide-ranging theories began to pop up. At this stage I knew very little of
the particulars of the PanCam so I decided to go and see what the Horses mouth had to
say. I sent out a swag of emails to the NASA marsrover team, the Athena Instrument
team at Cornell University, and the long shot, an email to Assoc. Professor James Bell.
Who is the *Pancam Payload Element Lead* for the mission.

Now, getting no response from the Athena team, and an automated response from the
NASA team. I was amazed and delighted to see that Dr. Bell had indeed taken the time
out of his busy schedule to help explain this quirk in the panorama pictures. His email
response is below:

> 66
> Thanks for writing. The answer is that the color chips on the sundial have different colors
> in the near-infrared range of Pancam filters. For example, the blue chip is dark near 600

The response regarding the image colouration read as follows:

*"Thanks for writing. The answer is that the color chips on the sundial have different
colors in the near-infrared range of Pancam filters. For example, the blue chip is
dark near 600 nm, where humans see red light, but is especially bright at 750 nm,
which is used as "red" for many Pancam images. So it appears pink in RGB
composites. We chose the pigments for the chips on purpose this way, so they
could provide different patterns of brightnesses regardless of which filters we
used. The details of the colors of the pigments are published in a paper I wrote in
the December issue of the Journal of Geophysical Research (Planets), in case
you want more details...*

*All of us tired folks on the team are really happy that so many people around the
world are following the mission and sending their support and encouragement...*

Thanks, Jim Bell Cornell U."

Did they also choose a particular pigment for the **connector clips?** (see the
previous image, showing the rover's colour calibration feature.)

Again, Dr Gil Levin weighed in on the issue with a paper entitled "Color
Calibration of Spirit and Opportunity Rover Images[318]." He concludes:

*Color calibration charts in production Mars Exploration Rover images should
either match the charts generated during calibration or should differ from them*

by a single uniform illumination model, expressed as overall multipliers for the red, green and blue channels.

Otherwise, production Martian images should either be made using the color chart to match Earth illumination, or should be made by trusting the luminosity calibrations made on Earth before launch.

In other words, "something still isn't right…"

"We're so pleased to be with you…"

In his eBook, "Fossil Hunters Guide to Mars," Charles Shults III has a section about the Spirit/Opportunity rover images entitled "Addressing the Issue of Color." In this section, he carefully analyses an image of the endurance crater and concludes that in the image that was released to the public, the sky has been "painted over".

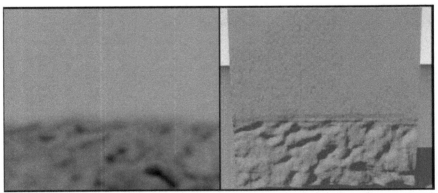

Figure 5.09 – Tthe NASA false color image shows where the sky was literally erased and painted over, creating the nearly flat artificial sky that we see (brown). They also blurred the sky into the ground to mate the two images together. On the right, the seam where the blend was done is apparent at the top of the rough crater edge.

Part of Charles Shults III's analysis of an image of Endurance Crater from Sol 097 – based on NASA/JPL image 1P136800137ESF2002P2559L2M1.JPG.[319]

In his analysis he concludes:

To make the two images fit together, the whole was blurred together and we see a great loss of detail as a result. The raw data is sharp and has plenty of contrast, the NASA painted sky image is clearly of inferior resolution. In examining the many sky images on the NASA/JPL web site, I found that virtually all had been painted over grossly. When I downloaded and assembled the raw data into a color image, this is what I got (in Figure 5.10). Analysis proves that NASA artists or image processors have altered the sky and ground to reflect some preference that they have been pushing for years- that Mars is somehow a very alien place. The truth is that Mars in most of its areas looks very much like Arizona or New Mexico. It would be very difficult for somebody to tell the two places apart when presented with pictures unless they were very familiar with the differences between the two.

He also writes:

> *[In] NASA's press release version of Burns Cliff, made using infrared as a color... The colors are so far off that it barely qualifies as a picture of Mars... There is no good technical or scientific reason for this representation.*

In the Hoagland/Bara "Dark Mission" book, the issue of incorrect colours in the Mars Rover images is also covered in some detail – and an additional point is made about the differing colour of the rover's airbags is raised – they are far too red in colour.

So why would NASA want to convince people that the sky on Mars has a reddish appearance – contradicting what the data captured (over a period of several decades) actually tell us?

Conclusions

NASA has consistently misrepresented the colour of the sky (and some surface features) and several independent experts have found this, working separately from each other. NASA's misrepresentation is not accidental. That means there is a reason for them hiding the true colour of the sky in most of the colour images from the rovers and landers. We will consider this issue again in chapter 17.

14. Phobos

ESA Mars Express Image of the Mars-Facing side of Phobos[320] - 22 August 2004

Phobos, discovered as late as 1877 by American astronomer Asaph Hall, is the inner and larger of Mars's two moons, the other moon (discovered at the same time) is Deimos.[321] Below, I will list some of the unusual facts about this satellite:

1.   A roughly egg-shaped body, Phobos is 26.6 km (16.5 miles) across at its widest point. The shape is very unusual compared to other moons in the Solar System – which are much more spherical.

2.   It orbits around Mars quite rapidly - every 7 hours 39 minutes – as far as I am aware, it is the only moon which goes around its host planet in a shorter time than the planet itself rotates. This means that when viewed from Mars, it rises in the West and sets in the East.

3.   It is very close to the planet, compared to our own moon for example. Its mean distance is 9,378 km (5,827 miles). This means that if you were in the polar regions of Mars, you wouldn't ever be able to see it – as the curvature of the planet is itself would obscure your view.

4.   The orbit is nearly circular and only 1° from Mars's equatorial plane. Again, this is unusual compared to other moons.

5.   Its orbit is actually decaying, very slowly, so it is expected to crash into Mars, or break up to leave a ring of fragments around the planet, some time within the next 100 million years.

More details can be found in a video by the Open University[322].

In 1945, American astronomer B. P. Sharpless made accurate observations of the motion of Phobos around Mars. It was calculations based primarily on these observations that led Soviet/Russian Astronomer Dr Iosif Shklovsky to propose, in 1959, that Phobos may be an artificial satellite. This was based on working out the mass of Phobos and the nature of its orbit. In a book called "Intelligent Life in the Universe[323]," which we will discuss in a moment, Shklovsky goes into much more detail.

Following Shklovsky's public statements, in 1960, Dr S. Fred Singer, special advisor to President Eisenhower on space developments, stated he agreed with Schklovsky, although Singer said the figures still had to be verified[324]. In a 1960 Astronautics newsletter, Singer suggested that if Phobos was hollow[325], its purpose would probably be to sweep up radiation in the Martian atmosphere, so that Martians could safely operate around their planet. Dr Singer also suggested that Phobos would make an ideal space base, both for Martians and earthlings. More recently, however, Singer seems more guarded. In a 2013 lecture for The Mars Society, he stated (at 8:36)[326]:

> *My first paper on the moons of Mars was published in 1968. That's 45 years ago and I speculated that they might be captured asteroids. That is obviously wrong. I don't believe this anymore, but somehow this found its way into the textbooks and if you google the moons of Mars, or if you look at a textbook, you'll find out that they're captured asteroids. They can't be captured asteroids - it's completely improbable, but then again, I don't know what they are or how they got there. We need to find out. It's one of the great puzzles in the Solar System. By the way, for purposes of politics, it's good to pretend that Phobos and Deimos are captured asteroids. Maybe we can persuade NASA to get little chunks of Phobos and Deimos, so we can examine them. So I'm willing to play along on this issue and for a moment I will forget that I'm against their origin as captured asteroids.*

Note that he is willing to play along with the idea they are captured asteroids to get NASA funding for further missions…

Singer is mentioned briefly in a 1963 article in NICAP's "The UFO Investigator" newsletter. (Jan-Feb edition, Vol 11, No. 7)[327] as is Raymond H. Wilson Jr., Chief of Applied Mathematics at NASA. The article reads:

> *NASA TO PROBE MARS MYSTERY MOON - Space probes are now being prepared to determine whether the Mars moon Phobos actually is a huge orbiting Space base, according to a high official of NASA (National Aeronautics and Space Administration.) The disclosure that NASA is seriously considering this possibility was made by Raymond H. Wilson, Jnr, Chief of Applied Mathematics at NASA, in a discussion with members of the Institute of Aerospace Sciences. Wilson revealed to the group that investigation of Phobos, long an enigma because of its strange orbit, is one of the main purposes of the Mars probes. Previously, several prominent astronomers had agreed that Phobos might be a gigantic orbiting space base launched long ago by an advanced race on Mars. But this is the first time the possibility has received official backing.*

After stating that Phobos might be artificial, Wilson said Mars probes would contain spectroscopic equipment to determine whether Phobos gives off absorption lines of aluminum. If so, he said, this would be a "most interesting discovery." The NASA official said that the Space Agency's decision to investigate Phobos was based on the fact that it goes around Mars faster than the planet turns on its axis. This could not occur naturally, he stated, according to accepted ideas of planetary formation. Phobos, he said, is the only satellite in the solar system to have a period of rotation shorter than that of its main body. (About one-third the time of Mars 25 - hour rotation.)

Three years later in 1966, the book called "Intelligent Life in the Universe[323]" was published. This was co-authored with Carl Sagan and, apparently, Sagan had only been asked to edit the book. When Sagan had finished adding all his viewpoints the book had doubled in length.[324]

Chapter 26 of this book has the title "Are the moons of Mars artificial satellites?" It is very interesting indeed. On page 363, Shklovsky highlights an unusual description of Mars' moons that was written down and published in 1726 – over 150 years before they were discovered. This was in the novel "Gulliver's Travels" by Jonathan Swift![328] Here is the relevant segment:

They [the Laputan astronomers] have likewise discovered two lesser stars, or 'satellites,' which revolve about Mars, whereof the innermost is distant from the centre of the primary planet exactly three of his diameters, and the outermost five; the former revolves in the space of ten hours, and the latter in twenty-one and an half; so that the squares of their periodical times are very near in the same proportion with the cubes of their distance from the centre of Mars, which evidently shows them to be governed by the same law of gravitation, that influences the other heavenly bodies . . .

These figures are remarkably close to what was later measured. As we mentioned earlier, Phobos' orbit is takes about 7.5 hours and it is about 9500 km from the surface of Mars – which would be about 16000 km from the centre of Mars. The diameter of Mars is about 6800 km – so that according to what Swift had written in his novel, Phobos' orbit is about 20400km from the centre of Mars. He was a bit out on the distance that Phobos orbits (by today's measurements – but maybe its orbit has decayed). How could he have guessed the correct number of (then invisible) moons, being much closer to the surface than, say, our moon?

Following the discussion of "Gulliver's Travels," Shklovsky goes into considerable detail about Phobos' orbit – even showing calculations involving calculus! He does this to show the informed reader that the body is almost certainly hollow.

Therefore, we are led to the possibility that Phobos—and possibly Deimos as well—may be artificial satellites of Mars. They would be artificial satellites on a scale surpassing the fondest dreams of contemporary rocket engineers.

Having shown that the orbit of Phobos is decaying, Schklovsky points out something else that is interesting.

This circumstance points out another difficulty in the assumption that Phobos has a natural origin, for it means that we are now observing Phobos during the last fraction of a percent of its lifetime, an unlikely, but not impossible coincidence.

Shklovsky also notes the reaction to his remarks originally made in 1959:

As soon as it was published, in the form of a newspaper interview, the hypothesis of the artificial origin of the moons of Mars became the subject of wide discussion. The majority of scientists were skeptical, a reaction which of course is completely understandable. However, not one scientific argument was advanced against the hypothesis. An article in the American press by the American astronomer G. M. Clemence, of the U.S. Naval Observatory, stated that the British astronomer G. A. Wilkins, who worked for some time at the Naval Observatory, had obtained results indicating that Sharpless' data were in error. In response to my inquiry, Wilkins indicated that no new results had been obtained concerning the motion of the moons of Mars. Thus, the assertion in the American press was repudiated by Wilkins himself.

So, it seems someone lied in an attempt to ridicule Shklovsky. Why would this be?

On page 376, Shklovsky concludes:

Even if future observations indicate that the reported secular acceleration of Phobos is spurious, the hypothesis that the moons of Mars are of artificial origin has nevertheless been provocative, and thereby has served a useful purpose. It reminds us that the activity of a highly developed society of intelligent beings could have cosmic significance and could produce artifacts which would outlive the civilizations that constructed them. This conclusion, as we shall see in the following chapters, has significant implications for the problem of intelligent life in the universe.

But what has happened since Shklovsky made all these informed remarks? What has modern technology revealed about the nature and appearance of Phobos?

The Viking Orbiter missions returned images of Phobos in 1977[329]. These images revealed the very unusual appearance of the moon, with linear and parallel grooves running along its surface. Clearer pictures have been returned from the more recent ESA Mars Express probe. Some of these images are shown below and in the next chapter.

Another ESA Mars Express Image Showing Phobos' grooves[330].

It is somewhat less easy to see that some of these grooves intersect at right angles.

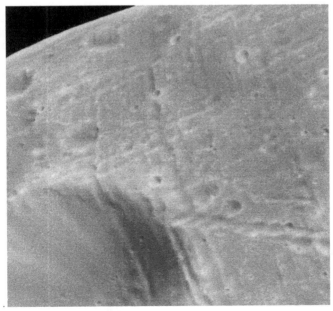

In close, we can see that some grooves intersect perpendicularly.

ESA's page about Phobos and Mars Express[331] contains some additional interesting information:

> *The origin of Phobos is being studied in detail but remains unclear. One idea is that Phobos is a captured asteroid. Data returned by the infrared mapping spectrometer experiment on board the Phobos 2 mission supported this view, but more recent investigations of its composition, including research with data provided from fly-bys by Mars Express, suggest that other processes may have been responsible.*
>
> *One reason to suspect that Phobos is not a captured asteroid is its density. Analysis of Mars Express radio science data gave new information about the mass of Phobos based on the gravitational attraction it exerts on the spacecraft. The team concluded that Phobos is likely to **contain large voids**, which makes it less likely to be a captured asteroid. Its composition and structural strength seem to be inconsistent with the capture scenario.*

This means that Phobos is hollow, in places… This research was discussed in Vol. 37, Issue 9 of "Geophysical Research Letters". In a letter entitled "Precise mass determination and the nature of Phobos[332]," the abstract, states, as indicated above:

> *We conclude that the interior of Phobos likely contains large voids. When applied to various hypotheses bearing on the origin of Phobos, these results are inconsistent with the proposition that Phobos is a captured asteroid.*

However, the ESA page continues:

> *It is possible that Phobos formed in situ at Mars, from ejecta from impacts on the Martian surface, or from the remnants of a previous moon which had formed from the Martian accretion disc and subsequently collided with a body from the asteroid belt. Data from the Mars Express OMEGA spectrometer suggests Phobos has a primitive composition, so primitive materials must have been available for accretion during its formation. The circular orbit suggests that Phobos formed in situ whilst analysis of the Planetary Fourier Spectrometer data from Mars Express also points towards in situ formation but does not rule out the possibility that Phobos is a captured achondrite-like meteor.*

Physicists would likely scratch their heads as to how a hollow object like Phobos would form by accretion (pieces sticking together) while it was orbiting the planet…

So, Shklovsky was correct - Phobos is at least partly hollow. Also, its appearance is unusual, as he suggested it would be, compared to other moons – it is not spherical.

Asteroid Vesta - Parallel Channels/Grooves

In 2011 and 2012, NASA's Dawn mission photographed Vesta - the second largest asteroid in the asteroid belt (approximately 25 times the size of Phobos) and it also has parallel channels. These channels run along its equator.

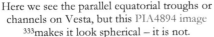

Here we see the parallel equatorial troughs or channels on Vesta, but this PIA4894 image [333]makes it look spherical – it is not.	Here is composite image PIA15678 [334]from the Dawn mission, which shows Vesta's shape more accurately.

Richard Hoagland claimed, in a 2011 article,[335] to be able to see geometric patterns in early images of Vesta. I have not been able to identify which images he was referring to and though there seem to be some hints of perpendicular lines [336]in some images, they don't seem anywhere near as clear as they are on Phobos.

Phobos Disinformation

In 1988, the Soviet Union, as it was then, launched two probes Phobos 1 and 2. Phobos 1 suffered a failure before it reached its destination[337] and so was unable to return any useful data about Phobos itself. Phobos 2 managed to get very close to the surface, but also suffered a mysterious failure, although it did manage to return some photos, which we will discuss below.

Three years after the demise of the Russian "Phobos" probes, in 1991, a story began to circulate about the "last photo" from Phobos 2. You will see this photo – and the story – still being circulated today.

The photo, seen below, was said to be a print of an infra-red photo. It was shown on the "Larry King Live" show on 22 November 1991.[338] Following this, an article by Don Ecker was published in "UFO Magazine" in 1992 alleging that the thin object at the bottom of the picture was a UFO – or even a missile. The Ecker article stated:

> *According to Zechariah Sitchin in his book Genesis Revisited, the rumor going around was that the Soviet spacecraft had encountered a huge "UFO" while in Mars orbit. In his book, Sitchin included a photo that the Russians released, which showed a large ellipse shadow reflected off Mars. Sitchin claims that the few photos Phobos sent back prior to disappearing were never released by the Russians, and that they treated the entire matter as "above top secret."*

We will look at this claim below. However, the article later continues:

The photo was given to Dr Popovich, she says, by Soviet Cosmonaut Leonov in June of 1989, and was reportedly taken by the infrared cameras on board the ill-fated Soviet craft. According to the Russians, the long ellipsoid-shaped object seen just outside of Phobos was 25 kilometers (approximately 15½ miles) in length.

UFO Magazine Vol. 7 Issue 01 1992

N E W S

A Soviet 'Close Encounter' By Don Ecker

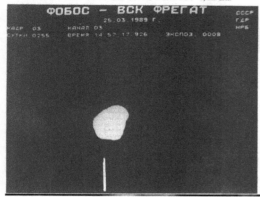

Left: Phobos 2 approaches the hollow moon… and a UFO is seen!!

The article then poses the question as to whether the object was a UFO of extraterrestrial origin. However, in this case, things become much clearer when you have more data. An article by "Mori" provides links[339] to more Phobos 2 mission data on the Planetary Society website[340].

Sitchin's book did not include the image of Phobos itself[341], (i.e. the one referenced in the Ecker article, above), but Sitchin references this image, suggesting that it shows something being fired at the Phobos 2 probe. Sitchin does, however, include the 2 photos shown below, as Plate O of his book:

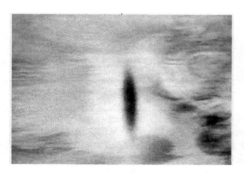
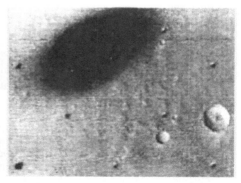

From Sitchin's "Genesis Revisited" Book: *"Mysterious" Shadows on the surface of Mars – Left (Plate N) is a still from a Soviet television clip and right (Plate O) is a Mariner 9 image*

Sitchin argues that these 2 images don't show a shadow of the same object. However, I think this is a very weak argument because the shape of the shadow would depend where Phobos was in its orbit and what time of day it was. The shape of the shadow would also depend on the slope of the ground that it fell on.

A Clearer View of the Phobos UFO

Below, we see 2 images from a sequence of about 15 from Phobos. The first image is a much clearer version of the one from the UFO magazine article. The second image shows us much more clearly that the strange object is just a digital artefact or "data drop out" of some kind. It isn't a UFO or a missile.

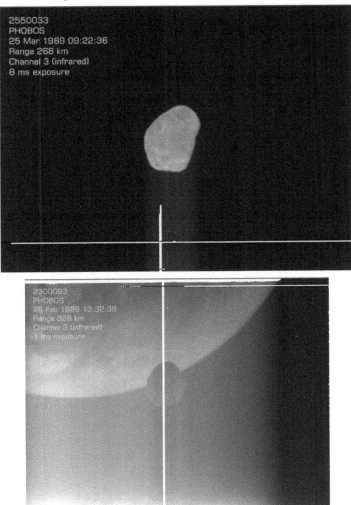

As regards the ellipse shaped shadow, this is almost certainly Phobos itself, as I alluded to – it has been photographed quite a few times now, as shown in this image from Mars Express

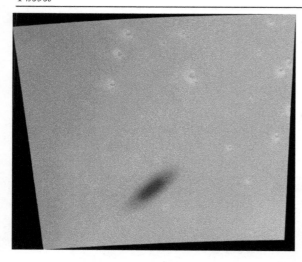

Left: An image from 2005 from Mars Express, showing the shadow of Phobos on the surface of Mars[342].

Conclusions

In this chapter, we have seen how very unusual Phobos is. We have also seen how disinformation has been circulated about it. I therefore tend to conclude, based on the evidence referenced in this chapter, Phobos is indeed, a partly or wholly artificial object.

In 2009, the story of Phobos took another strange turn. We will look at this in the next chapter.

15. Phobos, Monoliths and Buzz Aldrin

In 2008, some new high-resolution colour images of the strange moon Phobos were obtained by MRO. After these new images were released, a mixture of facts and disinformation was circulated, as we will discuss below.

The Phobos "Monolith"

The "flap" seemed to start with a report on C-Span – a US cable news channel. On 22 Jul 2009, they broadcast in interview/report with the tag line "Buzz Aldrin Reveals Existence of Monolith on Mars Moon."[343] In a peculiar interview, where a renewed programme of manned space exploration was being discussed, Aldrin then side-stepped unexpectedly, stating the following:

> *... if there's something very important to be developed from the moon... I'm not sure what it is right now and I think we should identify what it is for America to make such gross expenditures again for human habitation on the moon. We can help - we can join with, together... We can explore the moon and develop the moon. We should go boldly where man has not gone before... fly by the Comets, visit asteroids, visit the moon of Mars. There's a Monolith there - a very unusual structure, on this little potato shaped object that goes around Mars once in seven hours. When people find out about that they're gonna say, "Who put that there? Who put that there?" Well, the universe put it there. If you choose, "God put it there."*

No image of this monolith was actually shown in the C-Span report, though images of such a monolith had been discussed on some websites previously. It is not really clear to me what prompted this statement, other than possibly it was some type of media spin from NASA to obtain more funding for Mars missions. The "monolith" was initially discovered 11 years earlier (in 1998) by independent researchers Efrain Palermo[344] and Lan Fleming on a Mars Global Surveyor image[345] of the Mars-facing side of Phobos.

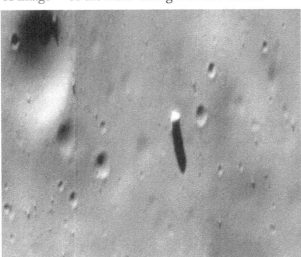

Left:
"Monolith"
discovered in
MGS image
55103h by
Efrain Palermo
in 1998[346].

In the image above, the "monolith" looks more like a dome, to me, but that might be due to the oblique angle and image projection that has been applied here. I decided to look at the original image on the MGS Website. I downloaded this and made a "flyover video"[347]. There seemed to be a number of objects which might qualify as "monoliths," although the one shown above is the biggest.

So why was this "monolith" being talked about by Buzz Aldrin, all of a sudden?

Aldrin Interviewed by Alex Jones

Alex Jones of infowars.com interviewed Aldrin on 17 Aug 2009[348]. Jones asked Aldrin about the Monolith and why it was interesting.

> *Jones: What do you think this is? I mean, tell me what your gut... or as a person - a doctor - who studies this... What does this look like to you?*
>
> *Aldrin: It's a big big tall rock. Now I can say it looks like maybe a crude construction device by some creatures - who practiced on Phobos and then landed in Egypt and built the pyramids. I don't really believe that, but some people are liable to think that.*

When quizzed about the "Egyptian" angle, Aldrin replied.

> *I'm pretty well convinced that that they were slaves that were conned into building this edifice for the Pharaoh, so the pharaohs could get to the other side in the Holy Land and put in a good word for all the slaves who built the pyramids.*

So why did Aldrin highlight the monolith on Phobos if he just thinks it is a "big tall rock?" He can't have it both ways, can he?

Alex Jones then describes something which doesn't seem to match the 1998 "monolith" image, shown above. We will further discuss this issue, below.

Media Muddle Up and Perception Management?

The Aldrin story was not reported accurately. Reports typically didn't include his full statement and, as is too often the case, talked about a "conspiracy theory" – when Aldrin himself did not use this term. The reports did not show the MRO and MGS image numbers. Details were wrong, and it seems a spurious and untraceable image was shown.

Shortly, I will show you a different "monolith" I discovered on the 2008 HiRISE images – none of the reports mentioned this particular artefact and nor were pictures shown of it.

The UK Daily Mail reported on 6 August 2009 "Buzz Aldrin stokes the mystery of the monolith on Mars" [349] and the report itself is rather jumbled and confused and includes the 1998 image, shown above, and another image, shown below.

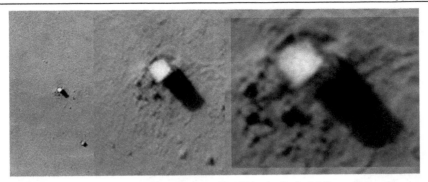

Image shown of the supposed "Mars Monolith"[350] in the Daily Mail – and elsewhere – in August 2009[349]. This image was attributed to an Italian website (see below).

The Daily Mail article attributed this set of 3 images as follows:

> *After being published on the website Lunar Explorer Italia, it set tongues wagging with space buffs questioning whether there was once life on the Red Planet.*

The article also talks about the image as follows:

> *How the experts see it: The original HiRISE satellite image supplied to Mail Online by the University of Arizona showing a close up of what appears to be a 'monolith' on Mars*

I have hunted for this image and cannot find it in the HiRISE archives or on Lunar Explorer Italia - or anywhere else. Some "Monolith" images are posted on the Italian site[351], but they describe it as a "rocky outcrop." The images shown on the Italian website do not give MRO or MGS or other image numbers and they are different from the ones shown in the Daily Mail article and they are different from the images shown in this chapter.

Hoagland's Phobos Monolith

A page on Richard Hoagland's Enterprise Mission website goes into great detail about Phobos and the "Monoliths"[352] and includes a reference image, showing where to look on the 2008 HiRISE/MRO images to see the Monolith (the new images are much higher resolution than the MGS ones and so it's hard to scroll around in a browser to find small objects). There are two new images numbered PIA10369 and PSP_007769_9010_IRB, which are from slightly different viewpoints.

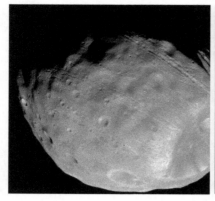 

MRO Image PIA10369[353]   MRO Image PSP_007769_9010_IRB[354]

Hoagland references the second image and shows you where to look for the "monolith."

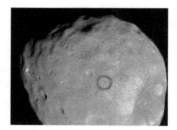

Phobos reference image from Richard C Hoagland's website[355], showing where to find the "Monolith."

Below, I have included two close up images of the region shown.

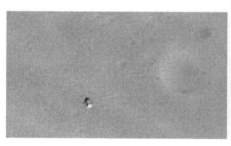 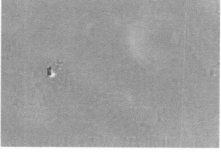

From MRO Image PIA10369[353]   From MRO Image PSP_007769_9010_IRB[354]

That's right… these images are actually *less* clear than the 1998 MGS images! So, why was there so much renewed interest in 2008?

"My" Phobos Monolith

When I was studying these images, I found a different "monolith." I think it's much more impressive! (Maybe I am biased!) See below for where to find it (or, if you're reading this on a device which can view video, you can watch a short video I made[356]).

Where to look for "my" Monolith in MRO Image PSP_007769_9010_IRB[354]

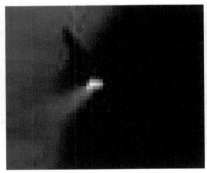

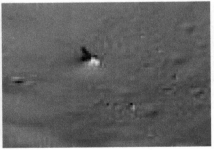

From MRO Image PIA10369[353]

From MRO Image
PSP_007769_9010_IRB[354]

As you will probably agree, this doesn't look that much like a monolith – it looks like a bright rocky-outcrop of some kind. I haven't been able to work out the size – maybe you can…

But now, let's have a look at something much more Monolith-like!

Valles Marineris Monolith – Close Up

Mixed in with the 2009 Phobos monoliths stories was an image taken by MRO/HiRISE on 24 July 2008. Remember, this is on <u>Mars</u>, not Phobos... A useful page on a German Mars-related website (marspages.eu)[357] provides us with the location of the "Monolith" in a wider image:

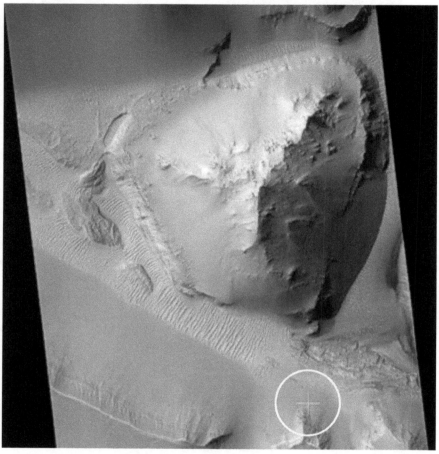

+ marks the spot where the monolith is located in PSP_009342_1725[358].

To see this, you will have to download the high resolution JP2 image and convert it to another format for easier viewing.

A very small section of the high-resolution image PSP_009342_1725 contains this:

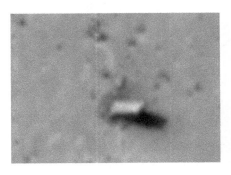 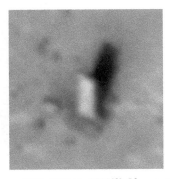

The "real" monolith? This is found in MRO image PSP_009342_1725[358]. If we turn this through 90 degrees, it really does look like a monolith! It's at location 7°14S, 92°37W, below a slope of "scree."

The area around the monolith has a couple of additional boulders or other sizeable objects:

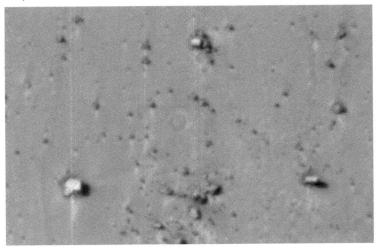

Area around the monolith. The 3 largest objects, when checked on Google Mars (they are *just* visible), seem to form something very close to an Isosceles triangle, with sides 74, 77 and 88 meters in length:

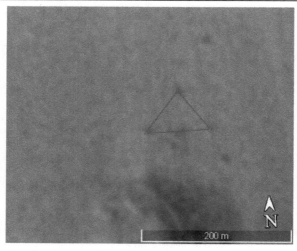

The base of this triangle runs within 2 degrees east-west – so this drawn triangle points north.

As we "know" however, it cannot be an artificial object, as an article in UK Daily Telegraph[359] advises us:

> *Alfred McEwen, professor of planetary science at the university and HiRISE's principal investigator, said: "Layering from rock deposition combined with tectonic fractures creates right-angle planes of weakness such that rectangular blocks tend to weather out and separate from the bedrock."*
>
> *He added: "It is not that unusual. There are lots of rectangular structures on Mars. "It is striking when you see one that is isolated, but they are common."*

Well, perhaps Professor Alfred McEwen has looked at more Mars images than me, but this "monolith" is the most rectangular-looking rock/boulder I have seen so far! The article continues:

> *Veteran astronaut **Buzz Aldrin recently stoked space conspiracy theory further** by announcing that a similar "monolith" had been detected on Mars's moon Phobos.*

Again, the article misses out most of the detail we have covered here and, as usual, "explains away" the anomaly with an "expert opinion" and a mention of "conspiracy theory." Also, doesn't it imply that the man who allegedly followed Neil Armstrong in walking on the moon is now a conspiracy theorist...?

The afore-mentioned "Mars pages" web page contains an excellent write-up about the monolith[357] and it shows you how to find it on the MGS image. It notes that the monolith is quite near a "Hill structure with a magnificent, pyramid-shaped mountain in the western Valles Marineris." Though it is "brave" enough to mention Arthur C. Clarke, "2001 a Space Odyssey" and so on, it generally "explains away" the anomaly and does not refer to any of the other evidence we have covered in this book. The text below is an automatic translation of what is written in German.

In a photograph of Mars Reconnaissance Orbiter from July 2008 from the area of the western Valles Marineris canyon system, a structure resembling the orbital photo is similar to the famous obelisk in Arthur C. Clarke's short story " The Sentinel " and the resulting By Stanley Kubrick produced film "2001: Space Odyssey", a milestone in the genre of science fiction film. Apart from all conspiracy theories of esoteric devotees, the NASA imagery is presented here so that the reader can form their own picture. Fig. 2 shows the overview picture in the western Valles Marineris at 7.2 ° S and 267.4 ° E. There are some larger hills and mountains in this area, which, together with a pronounced layer structure, characterize the bottom of the valley at this point.

The images show a uniformly shaped, cuboid monolith, as it also occurs on Earth, see for example here at Wikipedia. Monoliths are a product of the ice ages on Earth, which these uniformly shaped rocks have mostly deposited in end moraines. On Mars, this monolith is probably broken off the edge of the cliff and rolled down the slope, as shown by remnants of the fall trail of the rock. Such uniformly shaped stones are formed, for example, in magma deposits in the vent of a volcano, where after cooling such columnar structures can be formed.

This "monolith" story was reported seven years later, in 2016, in a BBC article[360], which is more accurate and gives more detail. However, the headline, I think, is misleading as it says, "There is a huge 'monolith' on Phobos, one of Mars's moons." As we have seen, the most monolithic object (which is mentioned in the article detail) is on Mars, *not* Phobos. The article doesn't give all the facts about Phobos, nor does it mention that the moon has "large voids," for example. Rather, the article refers to "the Face," in the following paragraph.

Perfectly natural erosional forces can also explain why Mars seems to be home to a levitating spoon and an Egyptian-style pyramid. Other spooky features, like the infamous face on Mars, do not seem quite so unusual when you take a closer look or view them from a different angle.

It's the first time I have seen the Face on Mars referred to as a "spooky feature." The article continues:

The Phobos monolith has not, as yet, received much scientific attention.

Why am I not surprised...?

Conclusions

I do think the Phobos/Mars monolith muddle-up story is yet another exercise in perception management. Things which indicate non-natural activity have been found on Mars and Phobos too – and we have been shown certain pieces of data which could prove this. Hence, figureheads like Aldrin are "wheeled out" to stir up the issue and cloud the truth, to the point that it is just too difficult or too time-consuming for people to find out the truth.

16. Mars Whistleblowers, Storytellers and Circadian Rhythms

Over the years, it is said that "whistleblowers" have come forward to tell us what they know about deep, dark and highly important secrets. In many cases, it is difficult to verify their stories. Oftentimes, I end up agreeing with the saying "those that know don't talk and those that talk don't know." In this chapter, I will look at what "whistleblowers" have said about hidden knowledge regarding Mars.

The Australian Cardinal George Pell and Mars

In an article in the Daily Mail entitled "The Pope condemns the climate change prophets of doom"[361] (13th Dec 2007), we read:

> *Australian Cardinal George Pell, the Archbishop of Sydney, caused an outcry when he noted that the atmospheric temperature of Mars had risen by 0.5 degrees Celsius. I must admit to being surprised at the use of the Industrial Military Complex as a phrase though...*
>
> *"The industrial-military complex up on Mars can't be blamed for that," he said in a criticism of Australian scientists who had claimed that carbon emissions would force temperatures on earth to rise by almost five degrees by 2070 unless drastic solutions were enforced.*

His use of the phrase "military industrial complex" is odd, in this context of referring to Mars.

Rush Limbaugh "There WAS Life on Mars..."

Rush Limbaugh is a news personality/commentator in the USA. On Mar 4th, 2004 he made a broadcast where he spoke of a secret "Gore Report" (initially he called it "The Mars Report")[362]

> *So it's clear that there was life elsewhere and now that NASA has kind of let the cat out of the bag, if there were habitable conditions up there we know there was life there. In fact, they found some DNA evidence up there that compares favorably to members of the Skull and Bones Society here on earth.*
>
> *In fact, they used the internal combustion engine. They were globalists; they got rid of all the countries on that planet and they had just one giant society, and the powerful, rich, just absorbed everything they needed from the best parts of the planet, used it for themselves and everybody basically suffered until everybody died out because what happened was there were no checks and balances.*
>
> *It's clear, ladies and gentlemen. Mars was once the jewel of the Solar System, and it was raped. Mars was raped. This is what NASA knows. This is what The Gore Report says. Mars was raped by robber barons. Capitalism, dependence on combustible engines led to global warming; the ecosystem that sustained everybody was destroyed.*

The whole segment in which Limbaugh says all this is rather peculiar. Why is he going into all this detail? Why is he clearly putting out disinformation – regarding the "Skull and Bones" Secret Society, which is surely made up of people from different families, all with different DNA. Why doesn't Limbaugh state where he got his information, or how the DNA samples were obtained. It's extremely weird. Is this a disclosure of some kind? Is it a disclosure that is so muddled up, no one would believe it? Was Limbaugh asked to make this speech because of what the Mars Rover missions might "discover"?

William Shatner (Captain Kirk)

In an informal interview on 9 April 2008, with a Q & A[363], when asked "Do you believe in Extra-terrestrial life?" Shatner stated[364]:

> *Of course, it is beyond a certainty that there is life out there. **I will let you in on a little secret, that I have been told not to reveal. So I won't reveal who told me, but there is going to be new information about Mars.** It won't be too long away.*

Clearly, he was not correct about the timing, or perhaps he was just referring to someone who was behind one of the many contradictory stories in the press, which have been circulating for the over 40 years now, which we discussed in chapter 10.

Disclosure Project Witness A.H.

According to Stephen Greer's disclosure project material, A.H. is a person who has gained significant information from inside the UFO/extraterrestrial programme groups within US government, military, and civilian companies. He has friends at the NSA, CIA, NASA, JPL, ONI, NRO, Area 51, the Air Force, Northrup, Boeing, and others. He used to work at Boeing as a surface technician. In an interview in December 2000[365], this anonymous witness stated[366]:

> *"I have another contact at NASA, JPL that I haven't mentioned. I can't mention too much about it because he is still working there. This person that I know is very high up in NASA. He said they know it's a face. They know that it was carved by somebody other than us. In an imaging area, they know for a fact that the face on Mars is real and that it was not made by windstorms or trick of lighting or anything like that. They know for a fact that the face on Mars was made by an extraterrestrial race that came here to Earth about 45,000 years BC."*

William Hamilton's Whistle-blower Source

At a Los Angeles MUFON meeting on 17 June 1998, UFO researcher William Hamilton[367] discussed meeting with an anonymous Colonel who had a degree in Plasma Physics and he had worked as an aerospace engineer. He was also a consultant to NASA and to people at the Palos Verdes nuclear power plant, 50 miles west of Phoenix Arizona. The former Colonel claimed to have worked at several underground facilities in the USA (having ridden to them on a Mag-lev train). One entrance to the system was at the White Sands proving grounds. The source told Hamilton that there was a base on the moon. Hamilton claims he was told by the Colonel that there was a secret space program and he had travelled on a craft that can leave the bounds of our atmosphere and travel to the planet Mars at 0.8c (eight tenths the speed of light). Hamilton said, "How do you know that?" The Colonel said, "I've been there." Hamilton continued:

> *I asked him "Where did you go? Where did you land?" and he said "We went underground..." On Mars there was a base there, where they landed and he would not describe the base - who put it there, or whether it was our base - nothing - he shut up about that. But in order to just evoke a little bit more information out of him, I asked the question "Are there Martians?" and there was a real pregnant pause before he replied, and he simply said "Yes." And I said "...and what do they look like?"... He said "like the ancient Egyptians." ... I think that's quite a coincidence because there were people wandering around in the 1950s here in California telling me they had met Martians and they had olive skin, dark hair, dark complexion - I don't know how many of those stories I heard. They were never little greys. They were always human."*

Storytellers?

The afore-mentioned anonymous whistleblowers seem somewhat credible to me, for several reasons – because they mention specific details (such as how long ago a race of beings came to earth from Mars, or how fast an alleged secret space craft can travel to Mars).

Over the years, a number of other people have come out and told stories about travelling to Mars, but for various reasons that would take too long to go into here, I have not found their stories credible. To me, the stories told by Andrew Basiago[368], Corey Goode[369] and Randy Cramer/Captain Kaye[370] tend to be a "gluing together" of various facts, rumours and myths and they present an overall picture which most people would laugh at or consider to be crazy. These people also seem to have been used to "suck people in" to subscription based podcasts and websites to either support them or their research.

Michael and Stephanie Relfe's accounts on "The Mars Records"[371], however, do strike me as slightly different and they have not tried to get people to support them financially nor have they regularly appeared on various talk shows and podcasts. On their website, they do seem to make some kind of effort to make informed comments on some of the matters I have discussed in this book.

Circadian Rhythms

In presentations that I have given in the past, I have mentioned the idea that if human beings are put in a controlled environment, but prevented from knowing the time – either from clocks, or by seeing anything to do with the usual day/night circle, they will revert to a 25-hour day[372]. If this is true, then one wonders if it has anything to do with the fact that the length of the Mars day is 24.9 hours. That is, did humans living on Mars get used to a day which was 25 hours long, before becoming used the 24-hour day that the Earth has?

Newer research in this area, however, suggests that if the human daily sleep/wake cycle is studied more carefully, and the subjects' access to the stimulus of light is controlled differently, then the natural cycle averages out at 24 hours 11 minutes[373].

I am given to wonder, however, what results would be obtained if different races of people were used in separate studies. I say this for two reasons – firstly, the origin of the races of humans is still something of a mystery. Could some of the races have a different planetary origin? Some years ago, when I started discussing these topics, somebody wrote to me who said that they knew an "insider" who knew more about the topics we have touched on in this book than we do. This "insider" said, in an "off-hand remark" that "white people are from Mars."

Conclusions

The information we have been given by whistleblowers is only partly useful – and is often mixed in with too much "other stuff" which is too difficult to verify. Having said that, I find what the A.H. whistle-blower said, and what William Hamilton's source said to be the most intriguing, and most easily "matched" with what I have come across in additional research.

17. The Apollo Hoax and the Rover Hypothesis

I must now include information that researchers Richard D Hall and Douglas Gibson began to bring to my attention in 2015[374]. Before covering that, we need to go to the troubling matter of the alleged Apollo Landings.

The Apollo Hoax – A Summary

Two researchers that I have already mentioned in earlier chapters are Richard C Hoagland[375] and David Percy[376]. In his book "Dark Mission" Hoagland maintains the Apollo missions happened more or less as NASA stated, yet he openly states NASA have lied about Mars missions and other data. We mentioned in chapter 3 that Hoagland contends Apollo astronauts were hypnotised to make them forget the "artefacts" they saw when they were on the moon. I contend that if they were "hypnotised," it was for another reason. I may well cover the Apollo Hoax in more depth in a later volume, although interested people can easily read Ralph Rene's "NASA Mooned America,"[377] Bill Kaysing's "We Never Went to the Moon"[378] or study David Percy's videos[379] and Jarrah White's videos[380]. Additionally, I have produced a presentation of my own called "Apollo – Removing Truth's Protective Layers[26]" and a short booklet based on this same presentation[381]. Another recommended video is "Moon Hoax Now" by Jet Wintzer[382].

Here, I present a few facts that should be checked by interested readers to help them understand that the Apollo missions did not happen as stated and perhaps the manned missions never even left low earth orbit.

- On July 20th, 1994, the 25th Anniversary of the supposed Moon Landings, Neil Armstrong addressed the crowd and said[383].

  "Today we have with us a group of students, among America's best. To you we say we have only completed a beginning. We leave you much that is undone. There are great ideas undiscovered, breakthroughs available to those who can remove one of the truth's protective layers. There are places to go beyond belief..."

- Photos of one foot of the LEM show little or no dust on them following the landing[384]. In general, the landing site shows no evidence of dust being blown around during the landing[385]. (There should be streaks/lines of dust stretching for many meters around the landing point.)

- Apollo 16 photo AS16-117-18841 allegedly shows a photo of Charlie Duke (and his family) left on the surface of the Moon in a plastic bag.[386] This shows something impossible - the plastic has not wrinkled/shrivelled <u>enough</u> when the surface of the moon in sunlight is over 100C+ heat[387]. (Try putting an old photograph in your oven.)

- Parallax studies of pairs of Apollo photos taken close together show that the distant landscape is a screen projection backdrop about 300 metres from the camera[388].

- The first LRO image of Apollo 11 landing site is extremely poor quality and does not show an object which resembles the Lunar Module closely enough[389].

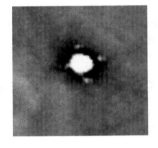

Other criticisms of the Apollo Hoax evidence suggest, for example, that reflectors left by Apollo astronauts prove they went there because we now bounce lasers off them. In response to this, I point people to an article in a 1966 National Geographic Magazine which reports that Lasers were being reflected back to the earth from the moon in experiments performed in 1962.[390] Further, a reflector was included on Lunokhod 1, which has successfully been used in more recent reflector experiments[391]. This means any such reflectors could have been remotely landed there – without any astronauts being involved.

Another common criticism is that rocks were brought back, which proves they went to the moon. However, it is quite well known that meteorites have been found in Antarctica which are chemically very similar to the alleged moon rocks. [392]Also, in 2009, a peculiar story was released on the BBC website with the headline "Fake Dutch 'moon rock' revealed."[393] This story stated that "A treasured piece at the Dutch national museum - a supposed moon rock from the first manned lunar landing - is nothing more than petrified wood."

Returning, for a moment, to "Dark Mission," what does Richard C Hoagland have to say about this sort of evidence, and one of the researchers that has presented it? On page 172 he writes (emphasis added):

> *As we discussed in the Introduction, this idea (most recently advanced by a few well-known **self-promoters** such as David Percy, Bill Kaysing and the late James Collier) had its origins as far back as the Apollo 11 mission itself. This*

myth is based on the simple—albeit **naive and absurd**—notion that the Apollo missions and subsequent Moon landings were "faked."

He continues:

> That said, one thing they [NASA] did not do, **unquestionably**, was fake the Moon landings. In fact, most of the charges made by these "Moon Hoax" advocates are so absurd, so easily discredited and so lacking in any kind of scientific analysis (and just plain common sense) that they give **legitimate conspiracy theories** (like ours) a bad name (which is more than likely now the real objective—see below). The comedy of errors and willful ignorance represented by the Moon Hoax advocates is too extensive to detail here. We instead refer our readers to the "Who Mourns for Apollo?" series on Mike Bara's Lunar Anomalies web site for a detailed, claim-by-claim dismantling of the whole Moon Hoax mythos.

I find it somewhat hypocritical of Hoagland to write in this manner – and he, then, has chosen to ignore evidence which would overturn his strongly-held beliefs. This is no different from what he accuses other people of doing. Also, Hoagland claims he has a "legitimate conspiracy theory." I think it would be far more sensible of him to refer to "evidence" than "theories." Those wishing to study Mike Bara's "Who Mourns for Apollo?" series can now find it on a blog he posted.[394]

Too Many Contradictions?

If we conclude that the Apollo Missions were faked, and that NASA has promoted an utterly false narrative, then how can we trust *any* of the photographic evidence I have referenced elsewhere in this book? This is a key question and one which has led some people to conclude the earth is flat – yes really! However, I can state with confidence that some of the observations about the planets can be verified from earth by amateurs. This takes NASA "out of the loop." Another consideration is that, for example, if the Cydonia images are fake, why would NASA and people like Michael Malin behave in the way they have done?

Rover Problems

Scepticism of authenticity of the Mars Rover and Lander missions to Mars is based on some technical issues, a few of which I will cover below. From one point of view, this could explain certain aspects of what we explored in chapters 8 - 13.

Before we look at the details of problems with the rover missions, I want to mention the NASA-assisted film that was released in 1977 in the USA (and released in the UK in 1979) – Capricorn One – which presents the story that NASA faked a manned mission to Mars and covered this up – killing 2 of the 3 astronauts that were (essentially) to be imprisoned for the duration of the mission[395]. Why would NASA support such a film? Perhaps I will cover more details about that in a future volume.

The Rover Hypothesis

This hypothesis, which was originally presented in 2014 by Richard D Hall and Douglas Gibson, was revised in June 2016. The hypothesis has been written up by Richard D Hall and a document containing the relevant data and evidence can be downloaded from his website.[396] Hall states the hypothesis as:

> *"The Mars exploration rovers are not situated on the surface of Mars, and never left the Earth"*

As Hall himself states:

> *The hypothesis may seem preposterous to anyone who is not familiar with the evidence contained within this document, therefore I would encourage readers to consider ALL of the evidence contained herein before dismissing the hypothesis.*

The main questions that the hypothesis asks are covered below.

Rover Batteries

How is it possible that Spirit and Opportunity rovers remained in operation for so long (over 6 years), when they are using Lithium Ion batteries, recharged by Solar Panels? We must consider that we need to renew laptop and similar batteries after much shorter periods than this. Additionally, the conditions on Mars are much more hostile than here on earth, which should have a significant effect on battery performance and battery longevity.

For the Curiosity rover, a radio isotope generator was used, but Hall is sceptical that this was tested sufficiently in suitable conditions to show that it worked. He notes that there is no soak testing documentation available for these rovers.

Wildlife Photographs?

Hall covers one particular photograph in detail - Curiosity Rover Image PIA16204[397].

On the left of this image, if we zoom in, we see this:

Left: A lemming? A Prairie Dog?

Hall compares this image to a Prairie Dog (found, for example, in Arizona and New Mexico) and a Lemming (found, for example, on Devon Island in Northern Canada).

Those not questioning NASA's story would, of course, suggest that this is just an unusual rock, seen from a certain angle. However, Hall spends some time analysing a second image - PIA16918 – of the same area, which suggests that this unusual rock "disappeared" later, indicating that this image does, indeed, show wildlife...

More Fossils and Plant Life?

Hall's document also includes two additional images I didn't include in earlier sections of this book – for example, Spirit Rover Image 2M160631572EFFA2K1P2936M2M1[398], which appears to show some type of Lichen.

Comparison of 2M160631572EFFA2K1P2936M2M1[398] with lichen

An additional image Hall includes is 1M131201538EFF0500P2933M2M1 from the Opportunity Rover Image. This is the Crinoid fossil image which was shown in chapter 11.

Explaining the Martian Sky Colour, Fossils, Plant and Wildlife...

Of course, it becomes obvious when you think about it... if the rovers weren't on Mars (and perhaps not the landers either), then we automatically have a reason why NASA would need to cover up the discovery of bacteria, moving microbes, fossils and so on... They couldn't possibly show the sky as being blue like, the earth... The same applies to "discoveries" of liquid water and mud.

Other Areas of Technical Difficulty

Hall notes that the landing mechanism (called Skycrane) does not seem to have been appropriately tested on earth. He suggests we watch a video about this.[399]

Hall also considers the Curiosity post-landing press conference, where Adam Seltzner (project leader) could not answer some basic questions about the project.[400]

Hall includes calculations relating to the descent of the Curiosity rover – and considers whether it was feasible that a parachute could have realistically been used to slow the lander down *enough*. The Martian atmosphere is *much* thinner than the earth's, though the gravity is less on Mars. On earth, a skydiver's parachute cannot be guaranteed to deploy properly above a height of 15,000 feet.[401] Hence, with a thinner Martian atmosphere, could the rover's parachute deploy correctly at 11,000 feet when the lander capsule was travelling 900mph?

In Hall's document, kinetic energy dissipation of the probe entering the Mars atmosphere (via the heatshield) and calculations relating to this are also considered.

The Mars Society and Mars Analogue Sites

Hall considers the role of the Mars Society and questions why this would appear to be independent of NASA, yet a key part of the Mars missions. Also, the possible role of "Mars analogue" sites is discussed, bearing in mind, for example, the "wildlife" photograph mentioned above.

MRO and the Curiosity Rover

There have been at least three photographs taken from Mars Reconnaissance Orbiter which are claimed to show evidence of the rovers. One claims to show the Curiosity Rover landing:

Curiosity landing – As Seen from MRO…?[402]

Another shows the rover on the surface, with tracks[403]. A further colour image shows Curiosity at another point on the surface.[404]

Are we really seeing genuine MRO images? Or, are they doctored/edited – as the LRO images of the Apollo landing sites must surely be. Sophisticated 3D modelling and mapping software is readily available, even to ordinary people now. You can, for example, run "Google Earth" in your browser on a computer with a decent specification and you can view detailed 3D models and maps of many cities and towns now. The Google Earth 3D model of my own house was surprisingly detailed. Could such software be being used to *simulate* the appearance of the rovers on Mars' surface?

Conclusions

It is kind of uncomfortable to consider that the Rovers and Landers are themselves a deception. I think some engineers might agree how difficult a controlled landing on a remote planet might be. I am now less sure about where the Rovers really are, though I am still convinced the orbiter missions are real.

18. Ringworld

Saturn is an incredible sight to behold even through a very basic telescope – the planet with its ring system has become a symbol of space and astronomy that we see as part of our everyday lives. Apart from its appearance and astronomical significance, some say that Saturn forms the basis for a kind of religion or a belief system, which world leaders are said to "pay attention to" as part of secret society activities. Much has been written about this, elsewhere, and it is a subject worthy of your attention. Carl James is one author who has tackled this complex subject.[405]

On 15th October 1997, the joint ESA/NASA Cassini mission was launched – it took almost seven years to reach the ringed planet. In 2004 and 2007, it photographed the strange "yin-yang" moon, Iapetus in considerable detail, which we will discuss in the next section.

The "active phase" of the Cassini mission lasted over 13 years, having come to an end on 15 September 2017, with the probe being directed to actually plunge into Saturn's atmosphere[406]. It seems to have been, by all accounts, a fantastically successful mission. Many books could be filled detailing the Cassini probe's scientific discoveries related to Saturn, its moons and its ring system.

Huygens Probe images surface of Titan[408] – 14 Jan 2005

In 2005, it even sent a lander down to the surface of Titan. The lander was named after the 17th-century Dutch mathematician, astronomer, and physicist Christiaan Huygens[407] – pronounced "hoyguns". Huygens discovered Titan in 1655. The Huygens probe returned a few pictures and much other scientific data. The image (left) has this accompanying description:

This coloured view, following processing to add reflection spectra data, gives an indication of the actual colour of the surface. Initially thought to be rocks or ice blocks, the rock-like objects are more pebble-sized. The two objects just below the middle of the image are about 15 centimetres (left) and 4 centimetres (centre) across respectively, at a distance of about 85 centimetres from Huygens. The surface is darker than originally expected, consisting of a mixture of water and hydrocarbon [methane] ice. There is also evidence of erosion at the base of these objects, indicating possible fluvial activity.

Much was learned, it seems, about Saturn's ring system and we will talk a little more about this shortly.

Saturn and Its Hexagon

First let us consider a discovery originally made by the Voyager probe, but later seen in more detail by the Cassini probe. This find wasn't that well publicised (certainly not in the 1980s reports about Voyager). It is the discovery of a long standing hexagonal feature at Saturn's North Pole.

This image hasn't been treated, for example, in the same way as images of Cydonia on Mars, or the images of fossils or trees, but the phenomenon is somewhat difficult to explain, though some people have made computer models to explain it.[409]

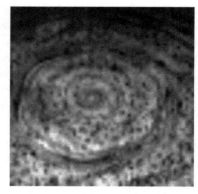

Thermal image of Saturn's North Pole taken in 2007[410]

Saturn's hexagon was photographed several times between 2004 and 2017 and, it has changed colour – from blue to yellow![411]

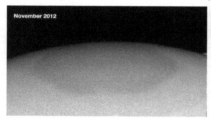

How can this peculiar feature persist? Perhaps we are seeing something related to the persistence of Jupiter's Great Red Spot (as we mentioned in chapter 6). Richard C Hoagland suggests that the hexagon appears as a result of torsion field or hyperdimensional physics, where rotation of a body has secondary physical effects. We can see these physical effects, but we cannot directly measure the energy which is causing them. It is interesting to consider that effects such as these do point to a set of physics which is not understood in the "white world" (but I contend that much more is known about this sort of physics in the black world). The Cassini mission has revealed many more mysteries for scientists[412]:

Puzzlingly, the rotation rate [of Saturn] that Cassini measured was slower than that measured 25 years earlier by the Voyager spacecraft, and that rate varied during Cassini's 13 years orbiting Saturn. Since an actual slowing of the giant planet's rotation was highly unlikely, scientists had a mystery on their hands. Various ideas have been proposed; some scientists suggested that material blasted into space by the geologic activity on Enceladus was to blame. Apparently, Saturn's magnetic field is slowed down as it drags through the ring of particles that litter the orbit of Enceladus. Other scientists have suggested that this strange signature originates in the upper atmosphere or ionosphere.

"Measurements from Cassini have totally changed our understanding of Saturn's magnetosphere, yet many questions still remain," said Burton.

I think these observations are mysterious to white world scientists because the physics they are using to explain the observations is incomplete - or just plain wrong - in some places. This is one of the areas where I give people like Hoagland and those that have informed his research due credit – they have tried to provide a better explanation for the observations and tie this in with other observed phenomena.

Another Hexagon…

Whilst on the subject of Polar Hexagons, I want to take a "quick jaunt" out to Neptune, before returning to Saturn. An Astrophotography specialist named Rolf Olsen has assembled an amazing digital composite of images of Neptune taken by Voyager 2 in 1989. He called his image "The Magnificent Neptunian System"[413]. It's best to view the original online, on a large screen to see the fine details – the faint rings of Neptune are shown.

Digital Composite of Voyager 2 Neptune Images

What was discovered in this image, but doesn't seem to be mentioned in Olsen's description is yet another polar hexagon formation in the clouds.

Ringmakers of Saturn?

Scientists have long puzzled over the formation of Saturn's rings and how they are structured. How long have they persisted? The usual explanation is, of course, to do with accretion and accumulation. However, a 1986 book called "Ringmakers of Saturn" by Norman R. Bergrun[414], describes a different idea. Bergrun, a former defence contractor, goes through his own analysis of Voyager photos using optical methods (enlargement). In the preface he writes:

> *Photographic enhancement has been accomplished by enlarging negatives with a microscope having recording and high-intensity lighting capabilities. Self-developing positive film recorded the various selected images contained in negatives. Copies of original photomicrographic recordings are the product of professional film-processing services.*

To my knowledge, it has not been updated based on more recent Cassini data, for example. Though the book has got good reviews on Amazon, the pictures in it are somewhat difficult to get excited about. Having reviewed a lecture given by Bergrun in February 2004, at the international UFO congress,[415] I remain unimpressed. Similarly, an interview he gave to Kerry Cassidy of "Project Camelot" in 2012[416] does not really give any useful information about how and why he is so confident about what he wrote in his 1986 book. Neither does he provide any updated explanations, unless I missed something. Instead, the 2012 interview covers many different topics, in an almost random (and unsatisfying) fashion.

Several people have asked me about Plate 5 in his book (reproduced below). Bergrun has annotated this photograph in some detail and, in a long explanation, claims that he can see "vehicles." Bergrun, based on an analysis of the photographs, concludes that these "vehicles" are thousands of kilometres in length. This is rather strange, because if these vehicles really were that big – having a length bigger than the diameter of moons that are visible from earth, why don't they appear more plainly in other observations? Bergrun does not seem to provide any other evidence that his vehicles exist, other than what he highlights in these photographs.

The images in his book do not seem to be sourced and, by today's standards are of poor quality. Again, to my knowledge, Bergrun has not referenced any newer images or data to back up his theories (compare this to what others have done with more up to date Mars data). He seems, to me, to make incredible extrapolations from a couple of grainy photographs.

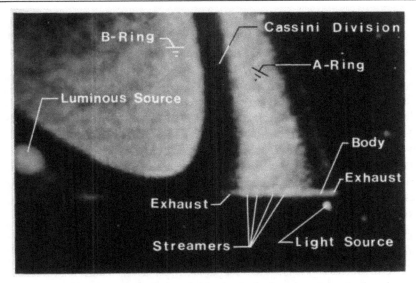

Plate 5: Efflux from along the length of a slender body, exhausting at both ends, generates the A-ring.

The source of the image above is not referenced (though it is assumed to be a Voyager Probe image), we cannot be sure what it actually shows. What we can be sure of is that it is contradicted by thousands of other images of Saturn – some of which have been taken by amateur astronomers, such as this one, from 2010[417], where we see a complete ring...

Photo by Anthony Wesley

Perhaps Bergrun's book is yet more disinformation – an attempt to discredit any future investigation of anomalies in Saturn's rings which might be more detailed or more thorough.

"Ringmakers" and Cassini UFOs

An interesting video I have shown in several presentations collects together some Cassini photo anomalies.[418] Pictures of these are interspersed with some of the images from Bergrun's book. It is quite an interesting video, but I really don't think it shows any proof of Bergrun's idea. Also, the video description says:

> *Norman Bergrun's "The Ringmakers of Saturn" is the inspiration for this video of the mysterious rings of Saturn. The big 3 amigo's on ATS (mikesingh, zorgon, and internos) have each added enormous effort of personal research on this subject, by gathering up to date Cassini satellite imagery that is showing some very interesting anomalies on and around the ringed planet.*

What "surprises" me is the people that have worked hard to find the images don't want to be identified by their real names. But then again, maybe it is not surprising – as the Above Top Secret forum is a place where debate is moulded and controlled by anonymous trolls and moderators (just like the vast majority of internet forums).

There are two image sequences shown in the video. The most interesting, to me, includes images N00084959 to N00084968:

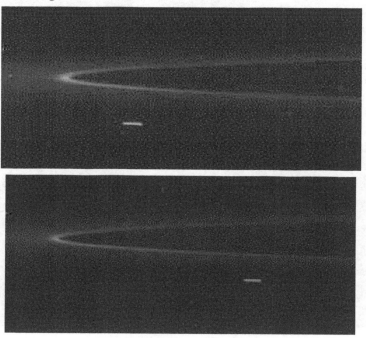

Anomaly in N00084960 [419] and N00084964 [420] (cropped and enlarged.)

A "something" is seen to travel across the frame – perhaps it is just a lone ice pebble, but the sequence is interesting. The image information states [421]

> *N00084958.jpg was taken on June 15, 2007 and received on Earth June 17, 2007. The camera was pointing toward SATURN at approximately 1,064,638 miles (1,713,369 kilometers) away, and the image was taken using the CL1 and CL2 filters. This image has not been validated or calibrated. A*

validated/calibrated image will be archived with the NASA Planetary Data System in 2008.

Another image sequence shows two bright objects moving down the frame. This could be just one of the other smaller moons, over-exposed. It does not look like a lens flare. If the probe is turning while these images are being captured, this could cause the apparent rapid movement of the object.

Two bright objects seen moving in an image sequence W00039344 [422]- W00039360[423]

The image information says[424]:

W00039344.jpg was taken on 2007-11-20 10:07 (PST) and received on Earth 2007-11-20 19:36 (PST). The camera was pointing toward Saturn- E ring, and the image was taken using the CL1 and CL2 filters. This image has not been validated or calibrated. A validated/calibrated image will be archived with the NASA Planetary Data System

Cassini UFO Disinformation

Having covered both the Cassini "UFOs" and Norman Bergrun's "Ringmakers" book, we will now briefly mention what I think was an attempt to draw attention to something which, at best, was inconclusive or at worst outright disinformation. On 28 October 2004 – Dr Steven Greer appeared on the Coast to Coast talk show programme. Below, I include the programme's synopsis, with my emphasis added:

*Greer said he was recently contacted by a **Lockheed Martin aerospace contractor**, who claimed to have direct knowledge of an official suppression of images sent to the Space Science Institute by Cassini. According to the witness, the Cassini probe had captured images of several large craft that appear to have been artificially created, and of "non-terrestrial origin." Curiously, the witness told Greer that the UFOs were not an "unexpected" discovery. Greer believes the government secrecy surrounding UFOs is largely based on their need to protect energy interests.*

This witness was not named – my guess is that it was... Norman Bergrun. His resume, which is included near the start of his "Ringmakers" book, states that he worked for the Lockheed Missile and Space Company in 1944-1956. This sounds too much like what is shown in "Ringmakers." Was Bergrun "jumping the gun" here? The images I included above were from 2007, but I find it too much of a coincidence to ignore.

Does this mean there is "something" floating around near Saturn that there was a chance Cassini imaging might pick up? So, someone wanted to put out a false story that could later be debunked? Perhaps only Steven Greer and Norman Bergrun know the answer to that question…

Vertical Structures in Saturn's Rings

In a 2009 presentation in Liverpool (see chapter 19 for more details), Richard Hoagland claimed that some type of artificial structure had been photographed in Saturn's B ring. The evidence he presented was basically non-existent – perhaps he was wanting to promote, too much, the Norman Bergrun "Ringmakers" disinformation. I include one of the pictures he showed here. It is, nevertheless, a fascinating photograph from the Cassini probe.

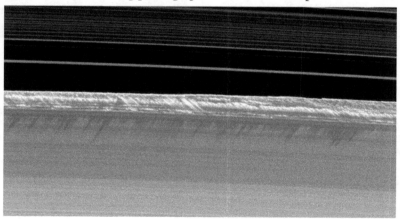

Image PIA116668 - Vertical Structures in Saturn's Ring cast shadows.[425]

Here is a slightly edited (for brevity) description of the image:

Vertical structures, among the tallest seen in Saturn's main rings, rise abruptly from the edge of Saturn's B ring to cast long shadows on the ring in this image taken by NASA's Cassini spacecraft two weeks before the planet's August 2009 equinox (26 July 2009). Part of the Cassini Division, between the B and the A rings, appears at the top of the image, showing ringlets in the inner division. In this image, Cassini's narrow angle camera captured a 1,200-kilometer-long (750-mile-long) section arcing along the outer edge of the B ring. Here, vertical structures tower as high as 2.5 kilometers (1.6 miles) above the plane of the rings... This image and others like it (see PIA11669) are only possible around the time of Saturn's equinox, which occurs ... every 15 Earth years. The ... sun's angle to the ring plane and causes structures jutting out of the plane to cast long shadows across the rings. This view looks toward the southern, sunlit side of the rings from about 32 degrees below the ring plane. Image scale is 2 kilometers (1 mile) per pixel.

Shepherd Moons

Whilst I doubt anyone would claim to know *exactly* how Saturn's rings formed and neither would they claim to know *exactly* how and why they persist, it seems that some clues were found in the data returned from Voyager 1 and Cassini.

Images from the Voyager 1 probe first showed evidence of what became known as "Shepherd Moons" – small moons that orbit near the rings and seem to "herd" ring particles as they travel around Saturn[426]. From the little that I have read, the "shepherding" process is complex and there is some argument about the details of what each moon does. It is said that "orbital resonances" are important in keeping the rings roughly as they are.[427]

It seems that eight of Saturn's moons are said to play a role in shaping the rings. We will look at two of those moons in more detail shortly.

Saturn's Innermost Moons: Pan, Daphnis, Atlas, Prometheus, Pandora, Epimetheus, Janus, And Mimas, to scale.

The information accompanying this composite image is below:

The eight innermost moons of Saturn, in color images collected by Cassini between June 7, 2005, and July 5, 2010. Pan and Daphnis (top left small moons) orbit within the Encke and Keeler gaps in the rings; Atlas (below Pan and Daphnis) orbits at the outer edge of the main rings. To their right are Prometheus and Pandora; Prometheus orbits just inside and Pandora just outside the F ring. Below them are Epimetheus (left) and Janus (right), which trade positions every four years, averaging out to the same distance from Saturn. Mimas orbits considerably farther away, but its gravitational effects influence the positions of gaps and waves within the rings. At full resolution, the montage has a scale of 500 meters per pixel.

Pan is about 10km in diameter and Mimas (the largest moon in the above photo) is about 200km in diameter.

Atlas

This moon was originally photographed by Voyager 1 and discovered by R. Terrile in 1980.[428] It is roughly the same size as Pan, and of a similar shape, though not so "hexagonal." Considering the title of the Norman Bergrun book and the fact that Atlas has been dubbed a "shepherd moon," I suppose there is a certain irony in the title of the article containing the image shown below - "Cassini Sees 'Flying-Saucer' Moon Atlas Up Close."

An early Cassini Image of Atlas – probably taken in 2007.[429]

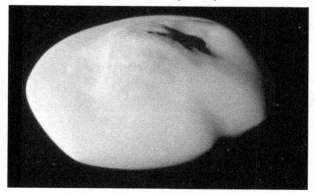

A close-up view of Atlas taken by Cassini [430] on 13 April 2017

Pan

This moon was not "fully discovered" until 1990, following image analysis by Mark Showalter. Its movements had been deduced from Voyager images, but it was Showalter who managed to identify it as a single pixel on one of the Voyager images.[431]

Pan is only 10km in diameter, so these are relatively small objects[432] (orders of magnitude smaller than Bergrun's "vehicles," for example). Cassini first showed, in 2006, that Pan had an unusual shape.

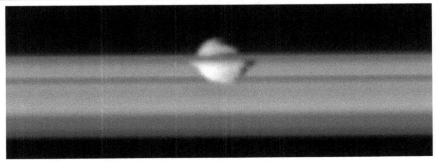

Cassini Pan Image from 2006[433] – Seen with rings edge on.

Much more detail was revealed in new Cassini Images from 9 March 2017[434]. Described by some as being shaped like a piece of ravioli[435], the reason why I wanted to mention it was because of yet another hint of it being a hexagonal shape.

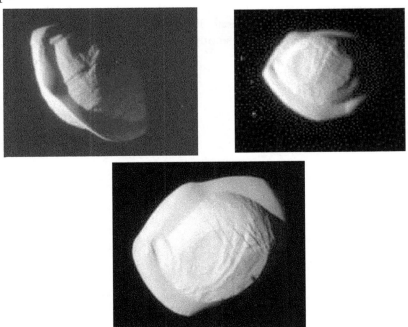

Cassini Images of Pan from March 2017

One explanation of the shape of this tiny moon is given by Planetary Scientist Carolyn Porco:

The leading idea is that the flange of ice around its equatorial bulge is ring material swept up and collected by the moon as it cruises through the Encke Gap. "Pan got its distinctive "skirt" because of the last stages of its formation (continued even in slow motion today) in which it accreted material from the rings it's embedded within," says Porco. "During the last stages the rings were

> *very thin and so the material falling onto Pan at this time came down on its equator and built the ridge you see."*

I hope she can explain the hexagonal (ish) shape too!

There is little more to say about these moons – I included them mainly because of their shape, and what was discussed in Bergrun's "Ringmakers" book. We can see they aren't "thousands of miles long." Are they just unusual accretions of rock particles, ice, dust etc, or are they ancient machines – operating on some unknown principles? Does the white-world physics explain the totality of their existence, their motion and their action?

Conclusions

The evidence in this chapter does not obviously indicate artificial constructs or the activity of extra-terrestrial life. However, the objects I have included are interesting in their own way and, I would argue, challenging to explain (despite what mainstream scientists would like to claim). I consider that the Bergrun book might be an example of deliberate disinformation, which was put out so that it can be debunked. We will see another example of this sort of activity in chapter 20.

19. Moon - Light / Shadow

In this chapter we will discuss Saturn's Moon Iapetus which is, like Mars's Moon Phobos, extremely unusual when compared to other moons in the Solar System.

The most important parts of the information in this chapter are available on the Cassini mission website. However, I must point out that I derived a lot of it, originally, from a series of articles written by Richard C Hoagland in 2004 – 2005. Whatever else I may have said about Richard Hoagland in this book and elsewhere, I really think he should put all the Iapetus information into a book, as the articles he posted at www.enterprisemission.com are most intriguing – and the subject matter does not seem to have been appropriately covered by anyone else. The title of his series is also apt - "A Moon with a View."[436]

Iapetus – the Yin/Yang Moon

In Greek mythology, Iapetus was a Titan – a God - the son of Uranus and Gaia and the ancestor of mankind. From the time it was discovered in 1671 by Giovanni Domenico Cassini, Iapetus has always been something of a mystery. One of the amazing things to consider is that Cassini observed Iapetus, over 740 million miles distant, using a 17th century refracting telescope whose main lens was only two inches in diameter! (A refracting telescope is one in which only lenses are used, rather than a combination of lenses and mirrors.) When Cassini observed Iapetus over a period of time, he was puzzled – every 40 days or so, it seemed to disappear from view, then re-appear 40 days later. He suggested that Iapetus had a light side and a dark side, and that it always kept the same face turned to Saturn (in the same way that our moon always keeps the same face turned to the Earth).

An Early Sketch of Saturn by Cassini

The orbit of Iapetus around Saturn is somewhat unusual – all but one of Saturn's other moons orbit "at the same level" as the Ring System (i.e. "in the same plane") – Iapetus orbits at an angle inclined to the ring system of about 15°. (Phoebe's orbit is inclined at about 5° in the opposite direction).

The reason for such an orbital inclination is generally assumed to be because the object has been "captured" by the gravitational attraction of the body that it orbits. In other words, it is thought that the object (Iapetus in this case) did not form out of the same cloud of material as the rest of the system. (Similar theories are used to explain the orbit of Pluto, which is also inclined to the plane of all the other planetary orbits). The angle of its orbit makes it the only large moon from which an observer standing on Iapetus could see the rings.

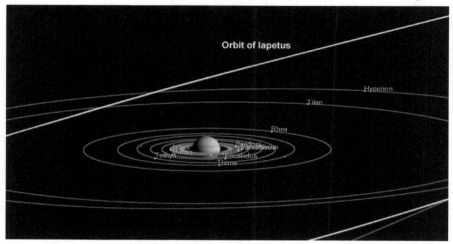

Orbits of Saturn's Moons (Size of Moons not to Scale)

Just over 300 years after the discovery of Iapetus - on August 20th, 1977, Voyager 2 blasted off from the Kennedy Space Centre at Cape Canaveral, Florida. A 4-year journey of fabulous discoveries took the probe to the Saturnian System. In August 1981, we got our first close-up view of this orbiting oddity. Not surprisingly, little fanfare was made of this event, because Iapetus is a relatively obscure object.

Iapetus - Voyager 2 Composite Image August 22, 1981[437]

As Cassini had predicted, Iapetus's surface was half dark and half light – the dark half was about 10 times darker than the light half. Saturn has tidally locked Iapetus, so it keeps the same face turned towards Saturn, throughout its 79-day orbit. This explains why, when observed from Earth, it appears to disappear for about half of its orbit. The Voyager image description contains the following discussion:

Amazingly, the dark material covers precisely the side of Iapetus that leads in the direction of orbital motion around Saturn (except for the poles), whereas the bright material occurs on the trailing hemisphere and at the poles. The bright terrain is made of dirty ice, and the dark terrain is surfaced by carbonaceous molecules, according to measurements made with Earth-based telescopes. Iapetus' dark hemisphere has been likened to tar or asphalt and is so dark that no details within this terrain were visible to Voyager 2. The bright icy hemisphere, likened to dirty snow, shows many large impact craters. The closest approach by Voyager 2 to Iapetus was a relatively distant 600,000 miles, so that our best images, such as this, have a resolution of about 12 miles. The dark material is made of organic substances, probably including poisonous cyano compounds such as frozen hydrogen cyanide polymers. Though we know a little about the dark terrain's chemical nature, we do not understand its origin.

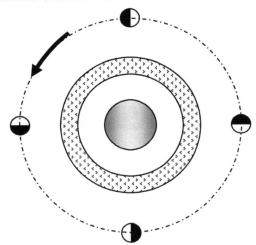

Iapetus orbits Saturn (not to Scale)

Voyager pictures seemed to show that the other features of the moon, however, were "run of the mill" – just consisting of cratered rocky features. The Voyager Spacecraft was used to make more accurate measurements of this unusual moon – to calculate its mass and density. Certain optical measurements were also made and these seemed to indicate that there was another potential mystery – the moon was not a perfect sphere – or at least, it was less spherical than it should be, when compared to other large Saturnian moons such as Titan and Enceladus. From a 2012 article by Emily Lakdawalla we learn[438]:

[Iapetus] is equatorially flattened by 4.5%, meaning that it's about 70 kilometers shorter pole-to-pole than it is wide across the equator. (For comparison, Earth's rotational flattening is only 0.3%.) This amount of equatorial flattening would make sense if Iapetus had a day only 16.5 hours long, but its days are actually 79 Earth days long.

The relative density of Iapetus was also calculated – and found to be about 1.1 or 1.2 (This means it is about 20 or 30% heavier than a similar sized body made of ice). The consensus of opinion is that Iapetus, and many of the moons of Saturn are mainly made of ice.

By July 2004, the Cassini probe had travelled to within 1.8 million miles of Iapetus. The picture it sent back showed the same elliptically-shaped dark region on the surface of Iapetus.

Early Cassini image of Iapetus – 3 July 2004[439]

Later, on 31st December2004, Cassini had a much closer encounter – passing at a distance of 40,000 miles.

Dark Side of the Moon

New high-resolution pictures were returned in late 2004 and posted on the JPL Cassini Mission Website on 7th Jan 2005.[440] They included an astounding discovery:

> *The most unique, and perhaps most remarkable feature discovered on Iapetus in Cassini images is a topographic ridge that **coincides almost exactly with the geographic equator**. The ridge is conspicuous in the picture as an approximately **20-kilometer wide (12 miles)** band that extends from the western (left) side of the disc almost to the day/night boundary on the right. On the left horizon, the peak of the ridge reaches at least 13 kilometers (8 miles) above the surrounding terrain. Along the roughly **1,300 kilometer (800 mile) length** over which it can be traced in this picture, it remains almost exactly parallel to the equator within a couple of degrees. The physical origin of the ridge has yet to be explained.*

Equatorial Ridge - Cassini Image PIA06166 Jan 2005[441]

We must also note that the Equatorial ridge *bisects* the dark region of Iapetus. In Richard C Hoagland's 2005 series of articles, he highlighted more (rather obvious) anomalies in the image above. I have not seen these referenced on any NASA or JPL web pages.

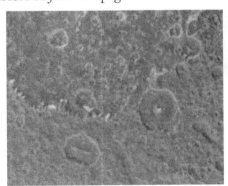

A closer look at the Iapetus images reveals some highly unusual craters. Two craters in particular are clearly not round. At least one crater is hexagonal in shape, with a raised mound in the centre. Another has an irregular shape (arguably more angular rather than circular) but it too, has an extraordinary linear ridge, roughly in the centre of the crater.

Whether this crater formed as a result of an impact or through volcanic action, how can a linear ridge (in the lower left crater) form approximately 24 miles (by my calculation) in length? Again, these extremely odd features are not addressed in NASA's main description of the photographs.

Hoagland points out more apparent hexagonal features:

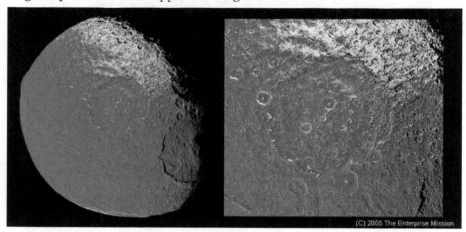

Hoagland notes additional hexagonal features on Iapetus' dark side.[442]

Another question we can ponder is that if the density of Iapetus indicates it is mainly composed of water ice, wouldn't this mean that the craters are more likely to be round, or at least, not hexagonal? Hoagland also highlights another possible oddity, in a Cassini image entitled "Seeing in the Dark" (PIA06146)[443]:

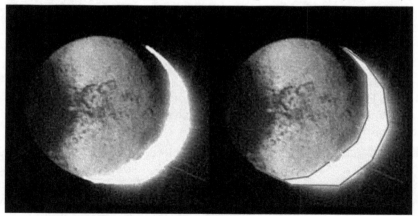

Hoagland notes a possible angular appearance to the moon's terminator in PIA06146.[442]

Is this just an imaging artefact? Having already seen hexagonal features on the moon, it seems harder to dismiss this observation. It also seems that a linear edge to the moon can actually be seen on some of the higher resolution 2005 images.

Ringworld NASA mini-documentary (2006)

Following the acquisition of many spectacular Saturn, moon and ring images, JPL/NASA produced a 30-minute documentary in 2006 called Ringworld. I was hoping this documentary would highlight all the unusual features of Iapetus. However, only about 1 minute was spent discussing the incredible moon[444].

> *Cassini has also had its first close look at Iapetus - a satellite with one face as black as asphalt and the other as white as newly fallen snow. Bisecting Iapetus' dark side Cassini has discovered an incredible ridge of ice that runs over 800 miles along the moon's equator and contains mountains three times as high as Everest. Elsewhere on Iapetus, Cassini's cameras capture the aftermath of a huge landslide that partially obliterates the floor of a Great Basin ringed by walls of ice over nine miles high. Later flybys of Iapetus will bring us views up to 100 times sharper.*

I found it surprising that they simply focus on a "landslide" feature and don't attempt to explain the wall feature and neither do they mention the hexagonal features – which are almost as obvious as the wall.

More on the Equatorial Wall

The "wall" is barely visible on the light side of the moon, as was shown in an image returned in 2007[445]. The image below also illustrates the enormous difference in colour between the 2 halves of the 900-mile diameter moon.

Light side of the moon Iapetus - Cassini Image PIA08384.[446]

Below are some of the closer 2007 images of the incredible "wall"[447]:

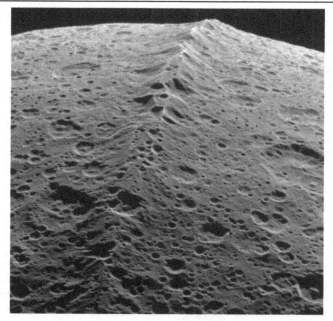

Image number W00035179.[448]

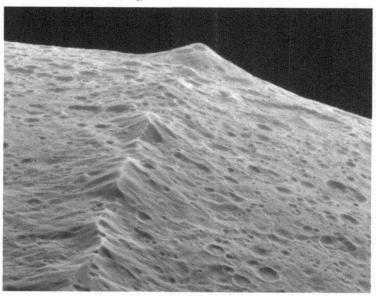

Image number W00035187.[449]

In the next image, we can see the gaps/discontinuities in the wall more clearly.

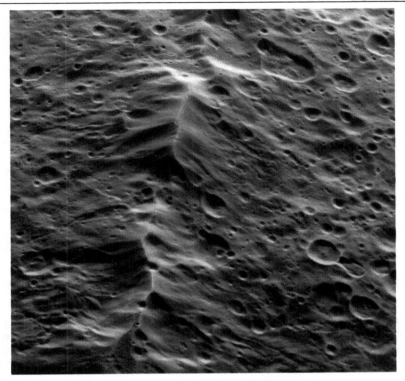

Image number W00035194.[450]

Formation of the Wall or Ridge

When the wall was discovered in 2005, these explanations were suggested for this extraordinary feature:

> *It is not yet clear whether the ridge is a mountain belt that has folded upward, or an extensional crack in the surface through which material from inside Iapetus erupted onto the surface and accumulated locally, forming the ridge.*

Another NASA posting read:

> *"The origin of Cassini Regio is a long-standing debate among scientists. One theory proposes that its dark material may have erupted onto Iapetus's icy surface from the interior. Another theory holds that the dark material represented accumulated debris ejected by impact events on dark, outer satellites of Saturn. Details of this Cassini image mosaic do not definitively rule out either of the theories."*

Since 2005, further explanations have been suggested. For example, an article on an academically oriented website "The Conversation" states[451]:

> *Due to its location and the very steep slope of its peaks it has been suggested that Iapetus may have once had its own ring of debris which has since collapsed down onto the moon, creating the ridge in the process.*

Well, it doesn't convince me! How would Iapetus have enough gravity to have a ring of debris form like this? How would such a ring form when the orbit of Iapetus is inclined at 15° to the plane of the rings? Sounds to me like ideas are just being thrown around. Maybe I am being unfair, or just dumb…

From the same 2012 article by Emily Lakdawalla that I referenced earlier, we learn more about the wall[438]:

- *The equatorial ridge is up to 20 kilometers high, 200 kilometers wide, and spans more than 75% of Iapetus' equator. If you do the math, you'll find it contains 0.1% of all of Iapetus' mass. This is both a little and a lot. All of Earth's oceanic crust amounts to 0.1% of Earth's mass.*

- *It is perfectly straight and sits exactly on the equator.*

- *It is heavily cratered, so appears ancient.*

- *Iapetus doesn't have any other major geomorphic features other than the ridge and lots of craters.*

The article makes another important observation about the ridge's ability to be able to "support its own weight".

…it's very surprising that a ridge so massive, so tall, and so narrow doesn't have a flexural trough running down each side of it. What that means is that the ridge formed at a time when Iapetus was so cold and stiff that the moon's crust has been strong enough to support the weight of that ridge without bending at all ever since the ridge formed.

The article refers to a scientific paper, published on 07 March 2012 called "Delayed formation of the equatorial ridge on Iapetus from a subsatellite created in a giant impact" co-authored by Andrew J. Dombard, Andrew F. Cheng, William B. McKinnon, and Jonathan P. Kay.[452] Building on their earlier work and that of other researchers, the paper attempts to explain the ridge's formation. The abstract for this paper states:

…we expand upon our previous proposal that the ridge ultimately formed from an ancient giant impact that produced a subsatellite around Iapetus. The orbit of this subsatellite would then decay, once Iapetus itself had despun due to tides raised by Saturn, until tidal forces from Iapetus tore the subsatellite apart. The resultant debris formed a transient ring around Iapetus, the material of which rained down on the surface to build the ridge. By sequestering the material in a subsatellite with a tidally evolving orbit, formation of the ridge is delayed, which increases the likelihood of preservation against the high-impact flux early in the Solar System's history and allows the ridge to form on thick, stiff lithosphere (heat flow likely <1 mW m^{-2}) required to support this massive load without apparent flexure. This mechanism thus explains the three critical observations.

On page 1 of the paper we learn a little more about the wall:

The ridge runs >75% of the circumference of the satellite, though not continuously, and has been modified by subsequent impacts and mass wasting (i.e., landslides). The cross-sectional shape is in places trapezoidal, with a flat top and sometimes a central trough, and with slopes of ~15°

For those that can't remember what "trapezoidal" means, it means "shaped like a trapezium!"

In the paper, it is noted that several models have been proposed for the formation of the wall. Some have suggested that the wall formed because Iapetus used to spin much more quickly – completing a revolution in only 10 hours rather than 79 days (190 times more slowly). This could have caused the flattened poles and the formation of the wall. However, the paper discounts this idea as the basis for the wall's formation as there are no other features which immediately show this is what happened.

The paper argues that because Iapetus' orbit is different to most other satellites, this makes conditions for the formation of the wall more favourable. They also note that there is a ring of "splotchy patches" on the moon Rhea, which is only about 1.8° from its equator and they therefore argue this is consistent with their hypothesis for Iapetus (although Rhea's orbit isn't inclined by 15° like that of Iapetus). Near the end of the paper, we read:

> *Fourth, the observed crater population of Iapetus should be studied. Based on cometary bombardment models [Dones et al., 2009], the age of the ridge should be estimated (i.e., how old is old?); this should be done in order to see whether the ridge is young enough to have avoided (or if older, could have survived) a Late Heavy Bombardment in a Nice-model-like rearrangement of the outer Solar System, and to test whether the apparently ancient nature of the ridge [Denk et al., 2010] is consistent with delayed formation.*

The age of the wall/ridge is a key issue. Regarding the craters, nowhere in the paper are the hexagonal ones mentioned. Also, the ridge feature is largely considered in isolation – the fact that it bisects the dark region is not – as far as I could make out – mentioned at all. Neither is the authors' own observation of the ridge's trapezoidal cross section discussed or explained. The implicit assumption seems to be made that we are dealing primarily with ice, but it is not clear to me what evidence they have regarding how much of the surface of Iapetus *is* just ice. How deep is the ice layer? They do refer to temperature changes in the moon's history, due to radio isotope decay – but it is hard to determine what data this is based on, without reading through several more scientific papers. Surely, it is key to understanding the ridge's formation to have a detailed understanding of Iapetus' structure and composition, otherwise it would seem that too much guesswork is involved.

Another "Trick of Light and Shadow"

As we have already mentioned, Iapetus presents another puzzle – the bright and dark appearance or "an extreme brightness dichotomy" to give it a fancier description. How can this be explained? In a December 2009 article "Global View of Iapetus' Dichotomy" on the JPL website, we can read[453]:

> *The cause of the extreme brightness dichotomy on Iapetus is likely to be thermal segregation of water ice on a global scale. Thermal effects are usually expected*

to act latitudinally. That is, polar areas are colder than equatorial terrain in most cases due to the more oblique angle of the solar irradiation. Therefore, an additional process is required to explain the longitudinal difference as well. In one model, dark, reddish dust coming in from space and preferentially deposited on the leading side forms a small, but crucial difference between the leading and trailing hemispheres, which is sufficient to allow the thermal effect to evaporate the water ice on the leading side completely, but only marginally on the trailing side. See Color Dichotomy on Iapetus to learn more. Iapetus' extremely slow rotation rate (1,904 hours), its distance from the sun, its relatively small size and surface gravity, and its outer position within the regular satellite system of Saturn are also crucial contributing conditions for this mechanism to work as observed.

In an article from "theconversation.com," referenced earlier, we can read[454]:

Iapetus and Phoebe orbit in opposite directions with Iapetus ploughing into debris ejected from Phoebe's surface. This debris forms an incredibly dark but giant outer ring around Saturn that follows Phoebe's orbit and is tilted relative to the planet's main rings. Cassini showed that the dust covering Iapetus slightly raises the temperature on that side so that ice cannot settle there. This means dark spots become darker, while water vapour transfers to the moon's other side making it even brighter. That's how Iapetus maintains its dichotomy.

My understanding of this is that they are saying that the moon Phoebe is associated with a rarefied cloud of material from which Iapetus is gradually collecting dust. So why would the pattern of bright/dark be like it is?

I think a far better explanation is that bright/dark dichotomy was formed when a cloud of sooty material hit Iapetus from some type of explosive event. The cloud of material passed by the moon in half of the time it took to rotate. Maybe this is too simplistic, but it seems to be quite a good fit with the Exploded Planet Hypothesis.

"Radar, Love"

Another of the instruments which the Cassini probe was equipped with is a sophisticated radar system, capable of several modes of operation. This instrument was used to good effect to map the surface of Titan (which has a thick orange coloured atmosphere mainly of Nitrogen and Methane, so the surface is hidden). Cassini's radar was used to obtain more data about Iapetus and its ridge's structure.

Iapetus had already been "radar scanned" from Earth in 2002 using the newly-upgraded Arecibo Radio Telescope and the results obtained were described by Gregory Black, of the University of Virginia:

It is known that the bright [trailing] side is mostly water ice, but we find it does not reflect the radar like other icy satellites that we've studied with the radar before. The ice on Iapetus appears much less reflective.

Over the course of the Cassini Mission, Radar scans of Iapetus were completed which, from my limited understanding, more or less confirmed the 2002

Arecibo results. Did the new radar studies reveal anything more about the structure of Iapetus?

A paper published in 2010 called "New Cassini RADAR results for Saturn's icy satellites" analyses radar results from the Cassini probe's scanning of several moons[455]. The language and results discussed in the paper's abstract are difficult to interpret, but it does say "Enceladus and Iapetus are the most interesting cases." Quite a bit of reference is made to concentrations of sub-surface ammonia being the cause of the particular radar profiles. It is also noted that at one radar wavelength, there is a correspondence between optical and radar reflectivity (albedo) of Iapetus. However, at the other radar wavelength, there isn't much correspondence.

For, Iapetus, the 2-cm albedo is strongly correlated with optical albedo: low for the optically dark, leading-side material and high for the optically bright, trailing-side material. However, Iapetus' 13-cm albedo values show no significant albedo dichotomy and are several times lower than 2-cm values, being indistinguishable from the weighted mean of 13-cm albedos for main-belt asteroids.

Getting a "plain English" answer regarding the significance of these findings would probably be quite difficult (so if anyone would care to oblige me...) The paper does not really seem to reveal much about the internal structure of Iapetus and neither does it reveal any further details which might explain the anomalous features of the moon (i.e. the bright/dark dichotomy, the wall and the hexagonal craters.)

Bergrun on Iapetus

Before moving on in our discussion of Iapetus, I would like to refer back to Bergrun's "Ringmakers" book, just to show that I am not ignoring it. I would also like to show why Bergrun doesn't help us solve the Iapetus mystery. Chapter 9 of the book is titled "Iapetus Mystery Unraveled." On Page 77, Bergrun discusses the colouration of Iapetus thus:

Observers of Iapetus have wondered how the iceous region, being shadowed from the sun, can be so intensely bright. They have wondered how the iceous surface can change so abruptly into a radically different asphaltic composition. They have wondered about unexpected flashes of light, large variations in surface reflectivity and sudden disappearances from view. The mystery is resolved completely and satisfactorily by the nearby presence of an active electromagnetic vehicle.

Earlier, on page 70, includes an exposure adjusted (but still unreferenced) photo of Iapetus, as shown below:

1. Cylindrical body
2. Nose end
3. Tongue
4. Isolated light source
5. Large light sources
6. Active zone
7. Quiescent zone
8. Underbody emissions
9. Radial links
10. Roll filaments
11. Roll filament source

Plate 38: Composite photograph of Iapetus showing illumination by, and a peripheral linking to, an electromagnetic vehicle.

Bergrun does not tell us how this image was created – what was the source image? Did he scan these images from prints himself? How does he know the dots "lights" aren't dust or scratches on the film, negative or print that he used? Speaking plainly, it seems to me like he has just seen random dots on his images and made something up to explain them. Yes, we are discussing issues of possible or even likely artificiality of Solar System artefacts in this book, but I cannot give Bergrun much credit here for careful cross-referenced or verifiable research.

Richard Hoagland on Iapetus – Can you see?

Though I found Hoagland's six "Moon With A View" articles on Iapetus to be quite fascinating, he seemed to follow his usual pattern of venturing a bit too far beyond what the evidence shows. An example is where Hoagland, in part 2[442], zooms in on one of the Iapetus photos (shown below) and suggests there are buildings on Iapetus:

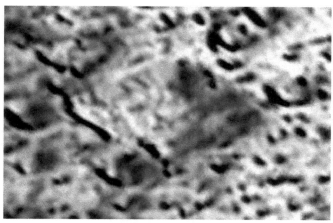

Does this image show buildings on Iapetus? (Which image did Hoagland use – and which part of it are we looking at?)

This is not something I would have stretched to, without additional evidence – it just looks like image noise to me. However, other aspects of the images that

Hoagland examines in this part may be more indications of artificiality, but the picture may not be as clear as Hoagland argues or implies. I think he has done similar things with the Iapetus data as he has with the Cydonia data.

Richard Hoagland on Iapetus in 2009

I attended a conference in Liverpool called "Beyond Knowledge" where I was asked to speak about the small amount of research I had done into the Crop Circle phenomenon. Richard Hoagland had also agreed to speak at this conference[456]. His presentation was called the "Secret Space Program"[457] The link here is a 35-minute selection from over 2 hours of material. I listened to the first hour, but had to leave the conference before Hoagland had finished his presentation. I did see him cover some of the Iapetus data, presented in this chapter, but Hoagland surprised me in that he did not talk about the hexagonal craters. Also, he barely, if at all, discussed the equatorial ridge/wall. Instead, he showed one or images of the alleged buildings on Iapetus and said something along the lines of "trust me, they are there." He skipped around and covered many different topics and included all kinds of things which diluted what he was saying.

Rigid Thinking

Since I originally wrote about Iapetus in 2005[458] (following Hoagland's interesting articles), more data has been gathered about Iapetus and more study has been carried out, on this extremely strange moon. I have referenced some of the published research about it. Again, however, I see rigid, straight-jacketed thinking in academia, which simply considers various combinations of geological processes to explain large-scale geometric features and other anomalies plainly seen on the moon. Going through this stylised, jargon filled research, it seems like the scientists have "everything pinned down." And yet, I for one, cannot accept the explanations proposed in academia – because they have still not fully explained the hexagonal craters and other features we have mentioned. Nowhere in academia is there any effort to consider Iapetus in a larger context. For example, I am not aware of any similar thinking to that demonstrated by Shklovsky regarding the nature of Phobos that we discussed in chapter 14. I believe that the data, even when taken in isolation, supports the idea that Iapetus is wholly or partly artificial in nature. Is it this conclusion which gives NASA such cold shivers that they simply ignore the most interesting data and hope that no one will notice? It is my view that when the data from Iapetus is considered in a wider context, it is even less safe to assume that it was formed entirely by natural processes.

With this in mind, and despite my earlier critique of Hoagland "overreaching himself," let us now consider some further ideas he proposed in his thought provoking series of Iapetus articles. We will also consider related ideas in chapter 22.

Richard Hoagland - A Bit of Informed Speculation

In his articles, Hoagland goes into further theorising about the orbit of Iapetus and how its distance from Saturn may be "hyperdimensionally significant." He also considers that certain other characteristics of the moon may indicate it is, indeed, an artificial object. He suggests that Iapetus has the shape of "a muted truncated icosahedron." He considers the materials from which Iapetus might be constructed. Another interesting idea he presents is that carbon-based materials could be used in its construction. Most people are already familiar with carbon fibre as a construction material – it has even been used for quite a few years in mundane items like bicycle wheels[459]. However, in 1985, a new type of carbon structure was discovered – and later synthesised. It became known as "Buckminster Fullerene"[460] as the structure of the molecule was similar to the geodesic dome created by Buckminster fuller. The carbon molecule consisted of 60 carbon atoms arranged in a truncated icosahedron. There are a whole set of related substances which have now been experimented with – including carbon nanotubes. Could a material like this be what helped to provide structural strength for the Iapetus' equatorial ridge? Is it being too simplistic to associate the colour of carbon (in the form of nanotubes or similar structures) with the dark colour of half of the moon?

Conclusions – Yes, "A Moon with A View of Saturn…"

As with Phobos, I think there are good reasons to conclude that Iapetus is a partly or wholly artificial object. I must say that the last of Hoagland's Iapetus articles[461] was the one I found the most thought-provoking and imaginative. It may be utterly wrong, but I found his suggestions quite captivating. The basis for his imaginative idea was already covered earlier in this chapter and I will repeat it here:

> *The angle of its orbit makes it the only large moon from which an observer standing on Iapetus could see the rings.*

Hoagland suggests that this as a reason why Iapetus is where it is – because of the view! He suggests it could have been a place of pilgrimage – perhaps like the Himalayas – where the scenery would inspire one's thinking. Iapetus would present a unique view of the Saturnian system – with an ever-changing array of moons and the incredible ring system. Hoagland presents some images of how this might appear and inspired by that, in 2005, I made a short video using the Redshift desktop planetarium programme. The video makes a short comparison between the view (over several days/weeks) from Titan compared to the view from Iapetus.[462]

I recently discovered there is free software called Celestia[463] which can also be used to create similar videos, with more detailed rendering.

Celestia – Simulated view from Titan – Rings always appear edge-on.

Celestia (Desktop Planetarium software) – Simulated view from Iapetus – Ringscape is ever changing.

The thing to notice here is that the view from Titan (and all the other large moons of Saturn) always sees the rings edge-on and the scene is much less interesting – you may not even really notice the rings of Saturn that much!

From Iapetus, we see a scene which changes every day and over a period of 80 days, we would see the Saturnian rings "opening and closing" while the numerous moons made their own rings around the planet. You would also be able to see, over the course of 1 orbit, both poles of Saturn. As Iapetus is further away than most of the other moons, they, too, are seen at different positions, rather than just in-line with the rings. Iapetus is, indeed, as Hoagland suggested "A Moon with a (breathtakingly stupendous) View."

20. Trouble in SOHO

In 2005, I became aware of another set of NASA/JPL imagery where anomalies had shown up. The particular example I saw seemed to be one of the most difficult to explain. I will start by showing the image and then explaining where it came from.

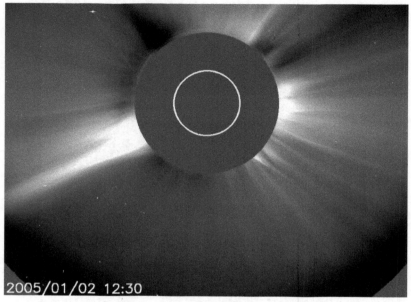

An image taken by the LASCO C2 camera on SOHO – 02 Jan 2005. A few copies can be found in the Internet Archive.[464]

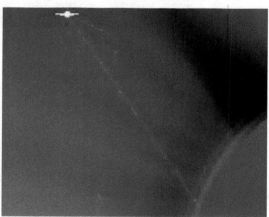

The interesting portion!

Knowing that this is a NASA/ESA image, directly downloaded from one of their websites, isn't it one that immediately makes you say, "What the heck is that??!!" Well, that was my reaction. The original image database has been moved, over the years, so the easiest place to see the image now is at the "Helioviewer Website." Go to https://helioviewer.org/ and enter the fields as shown in the screen-grab below (the Helioviewer images don't seem to have timestamps on them, however):

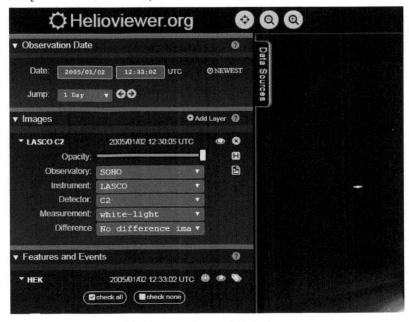

Helioviewer settings to find the image.

So, let us explain where it came from. This image came from the NASA/ESA Solar and Heliospheric Observatory – a probe that was launched in 1995 to study the Sun – and it is still in operation today (December 2017).

This mission is not widely known about – as the area of research that SOHO is concerned with is not of that much interest to people in general. Also, none of the pictures returned by SOHO resemble normal optical photographs, so many people would not understand what they are looking at if shown these pictures "cold."

SOHO has a number of instruments onboard, including two digital cameras C2 [465]and C3[466], which capture images mainly in the visible light range, but the cameras employ filters, so that only certain frequencies of light are captured. Each camera has a resolution of 1 megapixel and uses a different set of filters. The image above was taken using the LASCO C2 camera. The images from the C2 camera have a red or orange hue and the images from C3 have a blue-ish hue. The C3 camera has a much wider field of view than the C2 camera. The cameras are used for studying the sun's atmosphere, which is called the Corona[467]. Interestingly, these cameras have been managed by the US Naval Research Laboratory. (In chapter 14, we talked about Phobos being an artificial satellite. At the time, the one objection to the theory, proposed by Russian

Scientist Shklovsky, was published by a scientist from the Naval Research Observatory.)

The SOHO probe occupies a particular point about 1 million miles from the earth, where it has an uninterrupted view of the sun. It sits in the "First Lagrangian Point (L1), where the combined gravity of the Earth and Sun keep SOHO in an orbit locked to the Earth-Sun line."[468]

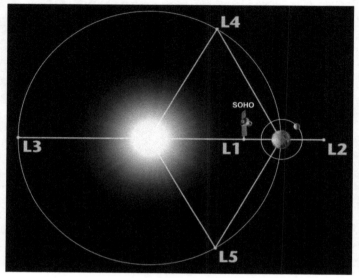

Graphic adapted from NASA Website.[469]

Returning to the anomalous image above, (that is, the one which appears to show some sort of craft firing beams at the sun), I used the Helioviewer to overlay C2 and C3 images from the same time. The object in the C2 image is right at the inner circle of the C3's field of view.

Left: Anomaly as shown on overlaid C2/C3 image.

Also, it is not clear if the C2 and C3 images are always captured at exactly the same time. Later, we will see more anomalies from both the C2 and C3 cameras. How does NASA/ESA explain these images?

Debunking SOHO Anomalies

On page entitled "How to Make Your Own UFO[470]" we find the following method listed. We will study the results displayed and try and relate it to the image I included above.

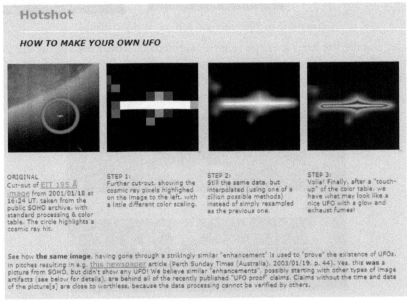

NASA/ESA's explanation for the anomalies.[470]

In Step 1, they filtered out the background noise, without being specific how it was done – what colour scaling was done? Down from how many to how many colours – 256 to 16 or 8?

In step 2, the explanation includes the words "(using one of a zillion possible methods)". Here is the definition of a zillion:

> *zillion zil¢yen, (colloquial) - noun an extremely large but unspecified number, many millions (analogous in formation and use to million and billion).*
>
> *(from Chambers English Dictionary, 1996 Edition)*

They, as scientists, should show more exactly what filter and how it might have been used. They could for example, choose one of those "zillions of possible methods" and try to show something more concrete. It also seems reasonable to suggest they may be able to give a more accurate or clear definition of filtering methods that may have been used to produce this effect such as "edge filtering combined with unsharp masking". The tone of the explanation is one of ridicule rather than trying to apply objective reasoning.

In steps 2 and 3, we can note that the object they show looks a little different to the one I showed above.

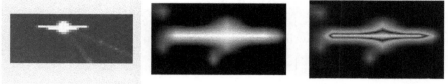

Comparison of 2005 anomaly (left) to NASA's "UFOs"

Now let us take the Step 1 image (which seems to be lower resolution than our object) and find out the proportion of the dimensions of the image artefact or object:

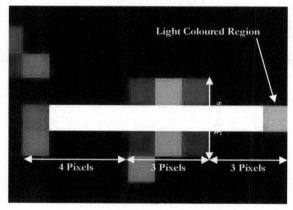

We can see that the ratio of the **height of the centre** to the **width** is

$$3:10 = 0.3$$

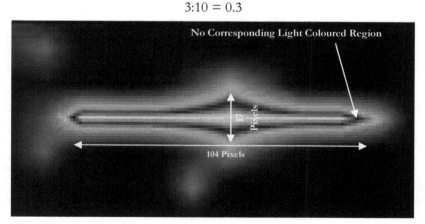

If we now look at the "Step 3" image, we can take the proportions of elements of this image. Measurement was done in *Paint Shop Pro* version 5, by selecting areas of the image and examining their size. The aspect ratio can be measured in several different places.

$$17:104 = 0.16 \text{ (measurements taken at edge of black halo)}$$

$$25:110 = 0.22 \text{ (measurements taken at edge of yellow halo)}$$

$$40:130 = 0.31 \text{ (measurements taken at edge of red halo)}$$

Whilst this isn't a firm measurement by which to gauge the final image, because it is unclear if an image filter has been applied, and what filter it is, the ratios do however, seem rather curious. Image filters should not change the proportions of an object in the image all that much (not ones which filter for noise anyway).

The correspondence of the first two ratios to that of the first image is not that good. Also, in the Step 1 image, there is a light-coloured pixel to the right. There is not really a corresponding light-coloured region at the right-hand side of the Step 2 or Step 3 image. Why did this get removed?

Notice in the NASA/ESA image, the background has changed colour between the Step 2 and Step 3 images.

Finally, there are other significantly clearer image artefacts in SOHO's image database which *have only been processed by NASA themselves*, so any image processing artefacts have also been introduced by NASA themselves.

They key point here is that the image I showed at the start of this chapter is **unfiltered** (except for any filtering that NASA themselves did *before* posting the image on their website). **Hence, their "explanation" is not applicable to the 2005/01/02 image.**

Also, the NASA explanation states: "Claims without the time and date of the picture[s] are close to worthless, because the data processing cannot be verified by others." The above picture has the date and time clearly imprinted in the image. It can also still be found in the SOHO archives, so their explanation does not apply to this image characteristic either.

Overall, the NASA image analysis and explanation of the anomalies is completely inadequate. However, this does not matter, because whenever anomalies like the 2005/01/02 image appear (and we will see more shortly), people inquiring about the anomalies will simply be directed to the "explanation" and no further comments from NASA – or anyone else – will ever be required.

Media Coverage

NASA's "How to Make Your Own UFO" debunk (above) also referenced a newspaper article from the Perth Sunday Times:

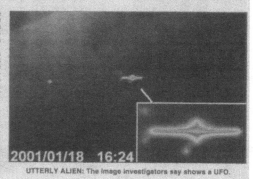

'UFO' on NASA camera

By TIM UPTON

WASHINGTON: The object is certainly unidentified and appears to be flying.

Whether this enlarged image really shows a UFO piloted by aliens remains to be seen. But according to the people who released it this photo and hundreds like it are the best evidence yet of the existence of spacecraft from other worlds.

UFO investigators say the image was captured by the Solar and Heliospheric Observatory (SOHO), a NASA satellite that was launched in 1996 to observe the sun. Since then, it is said, SOHO has captured hundreds of images of UFOs moving along a kind of alien superhighway.

SOHO is more than 1.5 million kilometres from Earth, with its camera trained towards the sun. Experts say the photographed objects are likely to be only hundreds of kilometres from its lenses.

Graham Birdsall, editor of *UFO* magazine, said: "The images are irrefutable in that they are from official satellites owned by NASA. They resemble the kind of spacecraft we used to see in sci-fi films like *Star Trek*."

2001/01/18 16:24

UTTERLY ALIEN: The image investigators say shows a UFO.

Story referenced by NASA/ESA SOHO anomalies "explanation" in Perth Sunday Times.[471]

The date of the story is not given, but it is presumably from January 2003, as similar stories appeared on the BBC Website[472] in the UK and the MSNBC website in the USA[473]. The latter story contains quotes from an ESA scientist:

> Last week, MSNBC.com forwarded press reports about the imagery to Paal Brekke, the European Space Agency's deputy project scientist for SOHO. In response, Brekke said he was aware of the claims and thought they were "quite funny." By Friday, Brekke and his colleagues had put together a more elaborate response.
>
> "Ever since launch, there's been a number of people who've projected their fantasies onto the SOHO images, seeing flying saucers and other esoteric objects," he noted. "Mostly, we're just amused by the unfounded claims, but in recent days, we've been receiving so many questions and claims (in news stories) that we'd like to set the record straight: We've never seen anything that even suggests that there are UFOs 'out there.' That is, to our (trained) eyes."

The pattern, then, is the same. They are certain they are *not* images of intelligently controlled craft. They then imply that to think they could be anything other than noise, debris, signal dropouts etc is just ignorant and foolish.

Neither the Perth Times nor the MSNBC stories name the source of the "saucer/UFO" image and neither explain how the silvery coloured object was created from the original SOHO image (which we see is dated 18 Jan 2001). The articles merely include a brief (and rather unhelpful) quote from UFO Magazine Editor, the late Graham Birdsall.

> Graham Birdsall, editor of UFO magazine, said: "The images are irrefutable in that they are from official satellites owned by NASA. They resemble the kind of spacecraft we used to see in sci-fi films like Star Trek."

However, the MSNBC article does include some useful additional information, about cosmic ray impressions, although most or all of this information is already included on the "How to make a UFO" page above. That is, NASA suggest that the object is not just the result of image filtering, but is caused by one or more of the following phenomena – Planets, Cosmic Rays, Software Glitches, Detector Defects or Debris. I don't think I need to point out how unlikely it is that such random effects would conspire to create the appearance of structured craft emitting beams in the direction of the sun...

More Media Coverage and Disinformation

The stories on the BBC website contain a little more information. The stories state that a selection of anomalous SOHO images was actually put on show at the UK's Leicester Based Space Centre in January 2003.

Stories as they appeared[472] on the BBC Website[474].

On researching this story, I was led to a press release dated 21 Jan 2003[475]. This press release was attached to the Leicester Space Centre[476], but I found it posted on a website of JSOC - Cluster Joint Science Operations Centre. [477]

This is located at the Rutherford Appleton Laboratory[478], itself part of the UK's Atomic Energy Research facility based at Harwell in Oxfordshire[479].

The press release makes for interesting reading and states that the SOHO images with the "UFO's" in them were presented by Mike Murray of a group called EuroSETI. (I have been researching the UFO subject since May 2003 - just after these images were discussed - and I have never heard of Mike Murray, nor have I heard of EuroSETI.) In the press release, it states:

> *EuroSETI's basic claim is that these images reveal UFOs near to our Sun. When asked why they are making these claims now, Mr Murray said, "With what we've got from the SOHO satellite, from our friends in Spain, it really seemed a good opportunity to show people, we believe for the first time, evidence of craft, that can only be craft, they can't argue that it's anything else."*

The press release also gives further details about the images which were given press coverage by the BBC etc:

> *Despite all the SOHO images being in the public domain, either via the Internet or on a free CD ROM, Mr Murray says their images are from a friend in Spain who is "spending a lot of time bringing down images from the SOHO satellite." The images EuroSETI are presenting are not in their original form, but have been digitally processed to 'enhance' them. The images Mr Murray is showing are not complete SOHO images, but small sections that have been enlarged. Every SOHO image is published with a time and date stamp for reference when used for research. We requested the time and date reference for some of the images, to enable us to put the small sections in the context of their original frame and pre-enhanced condition, but as yet we have not received this crucial information.*

Contrast the description of the EuroSETI supplied images with the image I included at the start of this chapter. The image I included is directly sourced from the ESA archive and has not been filtered or processed and it includes the date and time.

Next, we come to the disinformation promoted by the mysterious Mike Murray of "EuroSETI":

> *In an interview with UFO Magazine, Mr Murray claimed their images were from two different satellites, not just SOHO. He said, "...we looked at the LASCO satellite, which is in pretty much the same orbit as SOHO, but transmits in black and white, but the images were in that one as well."*

As the press release confirms, this statement by Murray is completely wrong. We covered, earlier in this chapter, how there are at least two cameras on SOHO. The ones we have referenced are C2 and C3 and cover a different field of view and different parts of the spectrum of visible light.

The rest of the press release mentions the "Cosmic Ray" explanation, but the images shown in the press release are all from C3 (a blue hue) and only one, partial, (and obviously filtered) anomalous image is included on page 2 of the press release.

To me, this has all the hallmarks of a managed disinformation promotion exercise – and it involved the Leicester Space Centre and someone from the Rutherford Appleton Laboratory. It is likely that the EuroSETI group was a front set up by the intelligence services to make sure no one got "the wrong idea" about the SOHO image anomalies. A (mostly) false version of the anomalies was presented to the press/media (and at the Leicester Space Centre) and then the false version (including errors) is debunked. Job done.

More Anomalies

Having discovered the anomalous "SOHO craft" image online somewhere, I posted it on my website some time in 2006 or 2007 and I had mentioned it a few times in presentations I had given. As a result of this, I was contacted by someone (who wanted to remain anonymous) in Germany who sent me a very detailed analysis of many more SOHO image anomalies that he had collated – all with times and dates. I posted his three-part analysis on my website[480].

Before the author of the analysis showed the anomalous images, he spent considerable time illustrating:

- Principles of the Visual Spectrum

- Principles of CCD of Operation

- Principles of Digital Imaging

- Principles of certain image filtering techniques

- Technical Details of SOHO imaging instruments

- List of explanations on SOHO website (Cosmic Rays, Debris etc)

This analysis, then, illustrates the author has a high level of technical competence and has studied and thought carefully about what he has compiled. He states:

- Analysis based on original NASA data files from SOHO near real time data archive and science archive:
    o   ~ 15,000+ C2/C3 images visual check
    o   ~ 1,000+ C2/C3 images filtered (5-10 filters)
- Used a random selection & filtering of 2006 C2/C3 archive (~ 2,000 – 3,000 images)
- Completed Image download & analysis from different web pages
- Completed Analysis of the available SOHO & instrument papers

Please study the analysis to learn more. I have posted a selection of about 230 images[481] from 2006 at http://www.checktheevidence.com/.

I have included a few of these images below, with a few brief comments or a brief description, and date labels, where these are not included in the image. These images have not been processed, only cropped to save space. You can probably find many or all of them via www.helioviewer.org. Following the image selection, you can read the author's conclusion about the meaning/implications of what he found in the images.

Anomaly Duration

One important point to mention is that the SOHO images are taken approximately every 30 minutes. Many of these anomalies appear on only one or two images at a given time. This then rules out that they are planets or comets – such objects do not travel so quickly across the field of view – especially with the C3 camera, which has a wider field of view.

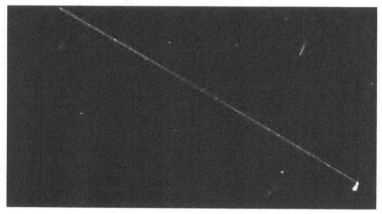

2006/04/24 23:06 – Beam with "noise" emanates from an anomaly - C2.

2006/05/14 18:30

Two beams and a small anomaly - C2.

2006/05/18, 06:30 - C2 (Cropped)

2006/05/25, 14:54 - C2 (Cropped and enlarged)

Beam emanates from bright object. (C3)

Two or three objects and beams at an angle to each other - C3

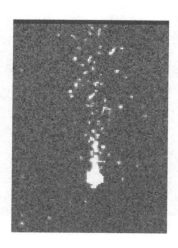

2006/02/04, 12: 18 – C3 – Beam from anomaly

2006/03/21, 12:42 – C3 – anomaly with strange "peppering".

2007/04/16 15:50

Beam emanates – towards Sun - from anomaly

2007/06/18 15:20

Multiple beams travelling in different directions – C3

The analysis concludes that some images have been altered and/or kept for longer before being posted online and that there have been other "shenanigans" with the images. The author includes the following remarks in his conclusion, after his lengthy analysis:

- The sun's activity is partly controlled by higher intelligence we do not understand yet

- Sun ejects massive amounts of energy partly triggered by higher intelligence

- NASA/ESA etc are involved in a massive cover-up and manipulate the images to hide the truth

- NASA is using image processing software to rework the images

- Coronal mass ejections might be the cause for the global climate changes we are currently faced with

- Some of the effects shown can't be explained at all

- CME's seem to cause shock waves hitting SOHO satellite

Conclusions

The conclusions given by the author of the SOHO analysis seem to be hard to accept, wild and "far out" and so on, but even if the author is only partly correct, it might go some way to explaining why the Naval Research Laboratory has been involved with the design and production of the LASCO instruments. (It is said by some that the US Office of Naval Intelligence is heavily involved in the UFO/ET cover up – and has been for decades.[482])

Some readers may already be aware that these SOHO images aren't the only space agency ones which seem to show evidence of structured, intelligently-controlled craft. They have "shown up" on several Space Shuttle missions – look for video from STS-48, STS-75 and STS-80, for example. A video analysis by David Sereda called "Evidence: The Case for NASA UFOs"[483] is particularly interesting.

As we have seen in earlier chapters, we have not only documented the wilful ignorance of the significant amount of evidence compiled from SOHO, we have also seen obvious disinformation circulated about it. Again, someone is hiding something.

21. Finding Planet X

This chapter is a little different from others in this book and is included due to the number of people that have written to me in the past about the alleged "Planet X." This planet goes under several names – Nibiru, Marduk, Wormwood and perhaps a couple of other names that I am not aware of.

A running theme in this book has been how NASA, ESA and other agencies or authorities concerned with Space and Space research have seemingly covered up, ridiculed or just downplayed significant data which indicates there is much more to our Solar System than they will admit. In short, they are covering things up. Could they also be covering up or denying the existence of a large planet in our Solar System? Is such a planet about to make a sudden appearance? We will now examine some of the speculation – and the evidence.

A Postulated Planet

Since at least the 1970s, there has been talk of a large, unseen planet being present somewhere in the Solar System. One interesting scenario that was presented in a 1969 science fiction film called "Journey to the Far Side of the Sun" (alternative title was "Doppelganger".)[484] The film's premise can be illustrated by reviewing the diagram from chapter 20. In the film, the idea was that the planet is orbiting the sun in the L3 Lagrange position, so is never visible from the earth.

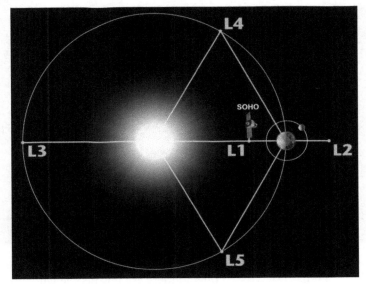

Since this science fiction film was made, a range of probes (Pioneer 10, Voyager etc) have travelled to points where they can observe the L3 point shown above and, of course, nothing has been found. More recently, in 2011, NASA's STEREO mission captured images of the far side of the sun[485]. This allows scientists to follow the progress of sunspots, and other features seen on the sun's surface, right around the disk of the sun (the sun rotates in approximately 28 days[486]) We can pause for a moment and note that the STEREO mission is also run by the US Naval Research Laboratory.

Planet X – Orbit

The Earth Chronicles books by Zechariah Sitchin[487], based on lengthy research into Sumerian Mythology as described in many cuneiform tablets and other writings, suggested there was another planet orbiting the sun. A single orbit by this planet took 3600 years. The planet, also called Marduk, was said to be home to the Anunnaki – a race of alien beings. The planet was said to be "almost the size of Jupiter or Saturn,"[488] (So if, as claimed, it travelled to the inner Solar System, it should be easily visible to the naked eye.)

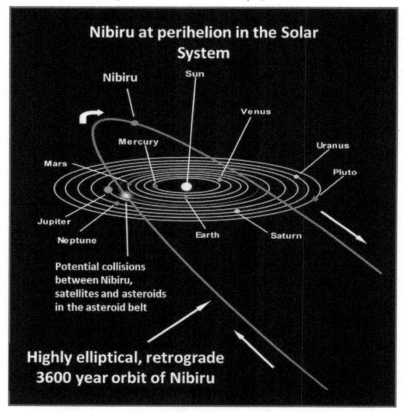

Proposed Orbit of Nibiru.[489] – **Please note the position of Jupiter.**

Zechariah Sitchin has now passed on and there are few people who would possess his knowledge of how to translate the Sumerian writings – and verify how he determined the nature of the orbit of Nibiru, as suggested in the diagram above, which was copied from the "Sitchin Studies" website.

Whilst I find Sitchin's work very interesting, and he does make some very interesting points, I have always felt uncomfortable with the idea of a large planet orbiting the sun as he describes. I have read enough about things like Kepler's laws, Bode's "law" and Newtonian Gravitation to realise what problems Sitchin's ideas present.

Observing Planet X

"Youtubers" and other internet commentators who have little or no knowledge of astronomy, and have never done their own research, do not seem to appreciate that there are people other than NASA employees who make important and valuable astronomical observations 365 days of the year. One example is in the search for comets and "minor planets." New comets are being discovered all the time – below I include a table I compiled from various sources of readily available astronomical information. Please study this in relation to the diagram, above, of Nibiru's suggested orbit and the relative positions of the other planets – particularly Jupiter. Also, please note that we need to get a good feel for the distances involved in what we will be discussing. We will be considering things in units of roughly **100 Million Miles**. This is also roughly **the distance between the Earth and the Sun**, so note:

1 AU = Distance from Earth to Sun = ~ 100,000,000 miles

So, for example:

> **Jupiter is about 5 AU from the Sun (between 4 and 6 AU from us)**

> **Saturn is about 9 AU from the Sun (between 8 and 10 AU from us)**

We can also stop to consider the relative sizes of the planets, which are shown in the pictogram below:

Relative Sizes of Planets.[490] (Graphic from IAU.org)

Here is a table of the actual measured sizes of the planets:

Planet	Diameter in miles	Planet	Diameter in miles
Mercury	3,032		
Venus	7,521	Jupiter	**88,846**
Earth	7,926	Saturn	**74,898**
Mars	4,222	Uranus	31,763

Note the sizes of Jupiter and Saturn, which anyone can observe for themselves, on a clear night, with a cheap/basic telescope – or even, now, with a superzoom camera[491]!

Remember, it is claimed that **Nibiru is Jupiter or Saturn-sized** – hence we can surmise it would have a diameter of about **80,000** miles.

We will now consider some of the comets that have been discovered over many decades and their **observed** size.

Comet's Most Common Name	Discovered By	Date Discovered	Discovery - Dist. from Earth (Million Miles)	Perihelion Date	Closest Distance to Earth (million miles) on last apparition	Size of Nucleus in Miles
Halley	Edmund Halley	1705	1000 (when re-observed on Oct 16, 1982)	09 Feb 1986	29	9 x 5
Hale-Bopp	Alan Hale / Thomas Bopp	23rd July 1995	670	01 Apr 1997	120	19 – 25
Wild 2 (pron. "Vilt 2")	Paul Wild	Jan 1978	~500	22 Feb 2010	62	2
ISON	Vitali Nevski and Artyom Novichonok	24 Sep 2012	590	26 Dec 2013	40	3

Note, again that these comets were discovered with (for the time) decent ground-based telescopes.

Let us take the example of Comet ISON, discovered by 2 amateur astronomers in 2012. When it was discovered, it was **590 million miles from the earth.** That is roughly the same distance away as Jupiter. Yet the size of the nucleus of comet ISON was only 3 miles in diameter. Even allowing for the glowing coma of a comet (which is not usually very significant when the comet is as far away as Jupiter), the size is four orders of magnitude (i.e. 10,000 times) **smaller** than Jupiter itself.

That is, if there was a Jupiter-sized planet moving through the Solar System, amateurs would have seen it long ago.

Dwarf or Minor Planets

When a new Dwarf Planet is discovered, or new research is published suggesting there may indeed be another planet in the Solar System, Planet X believers have been known to misquote and misrepresent the facts. A page on the "Northern Stars" website[492] lists some of the more recently discovered minor or dwarf planets:

QUAOAR is approximately 800 miles in diameter and is slightly farther away from the Sun than Pluto. It is made primarily of ices.

MAKEMAKE (pronounced MAH-keh MAH-keh) Diameter: 950 mi. (1500 km.), it's the third largest Trans-Neptunian Object after Eris and Pluto. Originally designated 2005 FY9, Makemake was discovered on March 31, 2005, near Easter, hence its original knickname of "Easter Bunny". Since Kuiper Belt objects are all officially named after creation gods, 2005FY9 was named Makemake for the creation god found in the traditional mythology of Rapa Nui or Easter Island.

ERIS (formerly known as 2003 UB 313 or Xena) was first photographed in October 2003, but was not detected until January 2005. Its discovery was announced in August 2005. This object is slightly smaller than Pluto and currently about three times farther away from the Sun than Pluto.. It is made of ices, much like Pluto. It takes this new planet 557 years to orbit the Sun just once. Its orbit is tilted nearly 45 degrees up at an angle compared to the plane of Earth's orbit. Eris has one known moon, Dysnomia

SEDNA is the most distant object yet discovered in the Solar System. It was discovered in 2003. It is roughly three times further away than Pluto, well outside the Kuiper Belt. It is approximately 1000 miles in diameter and takes nearly 10,500 years to go around the sun just once. Currently, scientists have not officially classified Sedna as either a planet nor a dwarf planet, Sedna was named after an Inuit Goddess of the Sea.

It is common to see talk of these objects being confused with Nibiru/Planet X – even though they are tiny in comparison to its alleged size.

Other stories appear from time to time about potential or theoretical discoveries of a larger planet, such as this one:

Huge 'Super Earth' planet ten times bigger than our own could be orbiting Sun

Left: Story from 27 March 2014 – UK Daily Express[493]

Nibiru's Orbit Is Not Viable

Planets orbit a star according to known laws, which have been studied and verified over a period of several centuries. Consideration of these laws in relation to the assumed or proposed orbit of Nibiru is discussed in an excellent article by Bruce McClure on the "Earth and Sky" Website[494]:

> *If it did exist, Nibiru's orbit would be highly unstable - Nibiru supposedly has an orbital period of 3,600 Earth-years. Let's pretend for a moment that's the case. Knowing the orbital period, we could use Kepler's third law of planetary motion to compute Nibiru's semi-major axis at 235 astronomical units (AU). Given the planet's semi-major axis, we could go on from there to figure out other aspects of Nibiru's orbit... If the semi-major axis = 235 astronomical units (AU), then the major axis = 470 AU. We know Nibiru's orbit around the sun can't be a perfect circle (orbital eccentricity = 0) because, if that were the case, it would always be 235 AU from the sun and never reach the inner Solar System. Assuming that Nibiru comes to within one astronomical unit (AU) of the sun, the outer edge of its orbit must recede as far as 469 AU from the sun... An orbit with such a high eccentricity is highly unstable... I find that at the Earth's distance from the sun, Nibiru would be flying at nearly 42.1 kilometers per second... Quite by coincidence, that 42.1 km/sec figure represents the escape velocity from our Solar System at a distance of one astronomical unit from the sun.*

Even if we wanted to say that Bruce McClure is just a "mainstream-thinking debunker" (which is not what comes through in his article), we still have the problem, mentioned above, that no amateur astronomers have photographed Nibiru/Planet X, while they have repeatedly photographed much smaller comets, such as Ison.

Planet X – Beliefs and Assertions

The above facts are difficult to argue with, but there are many who want to do just that. For example, in response to my pointing out the facts about the discovery of new comets by amateurs, they would say something along the lines of "Ah yes, but Planet X is dark – so it would not be seen easily." The darkness of such an object is not quite as big an issue as you might think. Any person with a basic knowledge of astronomy can tell you something about occultation – where one object passes in front of another, momentarily blocking out the light from it. For example, when the Moon passes in front of a star. Occultations are important for making astronomical measurements – indeed, it was how the rings of Uranus were discovered in 1977[495].

Hence, Planet X, if it was anywhere near where it has been claimed, would be observable by occultation, even if it was very dark. However, Sitchin said that Nibiru was a "radiant planet," which implies that it would not be totally dark[496].

Another common statement that comes up is "Planet X can only be seen from Antarctica – and a secret telescope there has picked up images." This is also a statement which ignores astronomy. There are very few areas which would only be visible from the South pole. In fact, only things very close to the earth – i.e. a few hundred miles above the South pole itself would only be visible from there. An object that was further away from the earth than a few hundred miles would also be visible from South America, for example. Also, the notion of a "secret powerful group having a telescope based at the south pole" is not well thought out. Powerful groups like most, or all, of the USA's three-letter agencies, will have access to much better technology – such as space-based telescopes – better and more sophisticated than the Hubble Space telescope. That is, they would have much higher quality information about a supposed "Planet X" than any ground-based observers would have.

Planet X - Disinformation

The topic has been rife with disinformation. For example, Nancy Lieder's Zeta Talk website[497] is quite extensive and she claims to channel "The Zetans" on demand to answer people's questions. However, some answers she gives have been hopelessly wrong... In 2002, the author of a "Planet X Critical" Website[498] wrote:

> *"Nancy started posting to astronomy newsgroups in 1995, about the time comet Hale Bopp was discovered[499], claiming that there was no comet. Hale Bopp, she claimed, was simply a nova used as a distraction, so people wouldn't see Planet X. In the spring of 1997 comet Hale Bopp put on a spectacular show even moving across the Orion area, very near where the mythical Planet X was supposed to be. Pretty strange for a distraction to move across the area it is supposed to be distracting people from! Now 7 years after this claim was made we still don't see Planet X, even without a distraction."*

Relating back to what we looked at in chapter 20, Lieder started talking about Planet X being visible on SOHO images in 2009/2010[500].

> *Planet X has been appearing on SOHO and the Stereo Ahead for some months now, in approximately the 4:30 o'clock position. But sometimes something bright with a tail shows up in other places, above the Sun, or to the left of the Sun. If this is Planet X, why is it moving around so much?*

This sort of coverage is poorly informed and has generally gotten more exposure than, for example, the analysis of SOHO images I referenced in chapter 20. Lieder has successfully associated the SOHO image anomalies with her bogus Planet X/Nibiru Story.

Another speaker who, I am ashamed to say, I took more seriously at one time is Jim McCanney. Mr. McCanney often calls himself "Professor McCanney" although his own biography [501] only shows that he was an introductory lecturer at Cornell University for 2 years in 1979-80. Nowhere does Mr. McCanney show that he earned the title Professor. He's apparently been working as a Certified Licensed massage therapist! He does not seem to have worked in a scientific or academic capacity since working at Cornell more than 20 years ago.

In 2006, he was a proponent of irreversible global warming.

A synopsis on the Coast to Coast Website of his 19 Mar 2006 appearance says:

> *"Far out in the Solar System, Planet X-type objects could become enshrouded in clouds of dust and gas and thus be hard to see, said McCanney, but he noted that a large object could be quite visible by the time it reached Pluto. Once in this position, such an object could take hundreds of years to approach Earth or it could come in at a "superspeed" in a matter of weeks, he warned."*

So, would Planet X travel slowly, or quickly? What would cause it to speed up? If he is such an expert in Physics and Maths, he should be able to state this more clearly. I will leave readers to investigate McCanney's own research and statements on their own, bearing in mind the simple and obvious points I laid out earlier in this chapter.

Planet X is Here!

I have, in the past, received several emails from people claiming to have seen Planet X. They have repeated the warnings about the cataclysmic effect it will have on the earth and the need, therefore, to prepare for this. I have asked them to send me a photo of Planet X and tell me more. Below, I include one of these emails, from someone who will remain anonymous.

> *"GG (21 Oct 2012): I took it from eastern Penobscot, Maine, looking west on October 2nd 2012 with a 12 mp Fuji digital and a green #12 filter. There are thousands of photos on poleshift.ning.com" "It came from the direction of Orion in 1983, came into the Solar System in 1999, came around from behind the sun in 2003 and stopped. As gravity pushes, it stopped the earth, venus, and the dark twin of the earth in its orbit while they stopped it in its orbit.*
>
> *Poleshift occurs in the 1 hour to break free of the dance/cups, and that will send a 600 ft tidal wave around the world, and in your case, UK will sink 150 ft.; therefore waves of 750-850 will erase the UK and Denmark. The date has never been announced."*

The photo this chap sent me appears to be a picture of the sun, taken through a green filter. The alleged Planet X he thinks he has photographed is just a lens flare. Photos showing lens flares are very commonly presented as evidence of the "elusive" Planet X being near the sun. For the moment, however, let me include more of this 2012 email exchange I had:

> *GG: "[Planet X is] 29,000 miles in diameter, cloudy oceanic, volcanic and atmospheric planet, 30 million miles from earth below the line from earth to the sun at the 5 o'clock position; its 5 million mile long tail pointed right at earth."*
>
> *AJ: So it's roughly 3.5 times bigger than the earth. And it's therefore about 7 times bigger than Mars. At 30,000,000 miles from us, it's about 1/3 the distance to the sun and 6 times nearer than Mars. So I guess it should be about 40 times bigger in the sky than Mars is (feel free to correct any of my figures). That's a pretty big object - and if it's greenish as your photo suggests, it should be pretty obvious.*

GG: *"And Zetatalk.com answers all your questions. It has its own google search engine for its own 39,000 pages of questions from the public and answers from the Zetas."*

Planet X Will Cause a Cataclysm!

There is a recurring theme that "Doomsday is nigh" – and predictions come around almost every year. The cause of "doom" varies, depending on what you read and listen to. Whole religions have been based on "believers" being saved either because they believe something or someone, or they have knowledge which allows them to better prepare for the impending chaos and destruction.

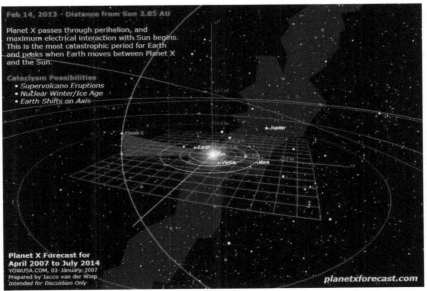

Feb 14, 2013 - Distance from Sun 2.85 AU

Planet X passes through perihelion, and maximum electrical interaction with Sun begins. This is the most catastrophic period for Earth and peaks when Earth moves between Planet X and the Sun.

Cataclysm Possibilities
• *Supervolcano Eruptions*
• *Nuclear Winter/Ice Age*
• *Earth Shifts on Axis*

Planet X Forecast for April 2007 to July 2014
YOWUSA.COM, 03-January-2007
Prepared by: Jacco van der Worp
Intended for Discussion Only

planetxforecast.com

Marshall Masters used to have a website called Planetxforecast.com [502]– this was a typical example of playing on people's fears of a disaster and how to prepare for it. Amazingly, even though he predicted Planet X's appearance in 2012 and he was of course, completely wrong, *he continues to promote the idea* in Jan 2018[503], as I am writing this chapter! Above is a graphic that was promoted by his website some 10 years ago.

Also in January 2018, the UK Daily Express has again carried a story about the appearance of Planet X entitled, "Nibiru SHOCK CLAIM: Planet X 'will arrive in May NEXT YEAR and trigger World War 3'"[504] The tabloid press once again demonstrates its consummate ability to promote false and misleading information, as it has done for decades.

For some further informed analysis regarding the Planet X/Nibiru issue, have a look at the "Cosmophobia" website[505].

Conclusions

I have seen no credible evidence that there is a large planet anywhere in our Solar System. It seems to me that if an amateur can detect a comet only 5 miles in diameter when it is as far away as Jupiter, they would easily see a larger object much further out – with advanced and semi-automated observing equipment.

All Planet X predictions have so far been hopelessly wrong and at least some of those promoting the stories don't understand the figures they are using and the images they are seeing – they don't seem able to compare them to data about the Solar System that we already know and can observe.

The Sun may have a binary star companion – or an additional planet – but it must be extremely difficult to detect if it is there bearing in mind that:

- Objects smaller than Pluto have recently been observed at far greater distances.

- Over 1000 extrasolar (outside the Solar System) planets have now been detected – by using missions like Kepler.

Though NASA could well be covering up knowledge about a present or ancient "Planet X", it would be far more difficult to keep Amateur Astronomers quiet about such things – they'd want their names in the history books!

Perhaps Sitchin's Planet X is not in the orbit he says/said – or has been confused in Mythology for something else.

It does seem like there may have been another Planet in our Solar System which exploded, as Tom Van Flandern (deceased) has hypothesised. Perhaps the promotion of "Planet X" in the media is a planned Psychological Operation to cover up something else and also to discredit other serious alternative research, such as what I have attempted to cover in the rest of this book.

22. Gatekeeping on Earth

I hope that if you have read this far, you have learned that there is powerful and important information that has either been downplayed, ignored or deliberately covered up. Have you been surprised by what you have seen and read? Have you been shocked at the idea that those tasked with researching and analysing data which would lead to some of the most important and world-changing discoveries have actually decided to keep secrets? How and why could this have happened? In this chapter and the following one, we will consider who some of the gatekeepers are and how they "keep the gates."

Creation of Consensus

A common theme in the different branches of research that I have done is that if you want to get people to look at new, compelling evidence, you usually run up against a consensus that will not want to "listen to what the evidence is saying." In astronomy and space research, the general consensus in established institutions can be summarised as follows:

- There is almost certainly no other intelligent life on any other planet in the Solar System.

- There probably never have been any other civilisations present in the Solar System older than our own. Our own civilisation has only been in existence for about 10 – 12, 000 years.

- There is no evidence that aliens or alien craft have ever visited the earth or anywhere else in the Solar System.

- There *may* be intelligent life elsewhere in the galaxy or the universe, but despite our best efforts, we haven't found any evidence of this.

- Governments and research and academic bodies and institutions do not cover anything up and have always encouraged free and open discussion of topics such as extraterrestrial life. To suggest otherwise is a result of paranoia and a desire to "believe in something" without any real evidence.

- There may have been some cataclysm in the Solar System long, long ago, but there isn't much evidence to tell us what happened and whatever it was probably wasn't that significant really.

Whether this consensus will ever change depends on whether the power structures that run the planet will ever collapse. The power structures are concerned with maintaining a consensus about many different things – health issues, banking and money, energy and energy technology, human origins and the nature of human consciousness and related matters.

In all the areas where a consensus view needs to be maintained, authority figures are present. In most cases, to become an "authority figure," it is necessary to follow a certain path – to "jump through certain hoops." The process of jumping through certain hoops forces acceptance of certain concepts as being true and you are forced to do things in a certain way, according to accepted norms. If you do not "conform to the norm" you will likely be thrown off the path and will be prevented from becoming an "authority figure." Hence, if you have important things to say, it is less likely you would ever get an audience of any significant size.

We can see some examples of "conforming to the norm" – in chapter 4 – papers about the Cydonia region of Mars, suggesting that it contained "archaeology" were much more difficult to get published – because the implications of what they showed went against the consensus view.

In chapter 19, we saw how the studies of the "wall" on Iapetus only consider geological explanations – considering anything related to construction or artificiality would go against the consensus, so it is likely papers analysing evidence for this would never be published in an establishment journal.

This book is being self-published because it goes against the consensus view.

How Consensus Thinking Hampers Research

Around the time I "found" the dome and "T" images, in 2004, I was following the ill-fated Beagle 2 mission and attended a talk by one of the scientists involved with the mission – Dr Geraint Morgan[506].

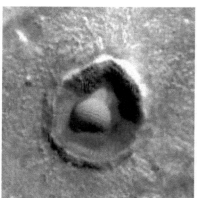 

I e-mailed him to ask him about these images and I included the links to the USGS/MSSS websites. He said this wasn't his area of expertise and he then passed them on to one of the other mission team members. (I did find it odd that, as someone who worked on the Mars Beagle 2 mission himself, he did not seem to be curious about these images.) At the time, I only highlighted the "crater dome" image and the "T" image SP243004, shown in chapter 6.

I asked could they explain the features of the Dome Structure, or even provide a model (briefly) for how it may have formed. First, I got this response:

"They look like longitudinal dunes to me. It would be nice to get the topography, but Mars Express will only have about one third of the resolution at this latitude."

Then I got this response:

"...these are natural features, as with 'the face', which was subsequently shown to be a collection of hills. The human mind is of course pre-programmed by evolution to identify patterns and assimilate them in to 'pictures' we recognise. More data is required of such features to fully identify."

Here, again, we see precisely the same pattern of qualified scientists taking little or no interest in this data. We have another instance of the scientist trying to compare it to "the face" and write it off as a trick of light and shadow. This feature looks nothing like the face!

It seemed to me that there was not much information in these responses, so I persevered a little more. Again, in 2004, I sent these pictures to 42 geologists picked at random from University Geology departments in the USA and the UK. 3 of them actually were kind enough to respond. This response came from an Associate Professor of Geology at an American University:

These two images from the surface of Mars are certainly puzzling and have been floating around UFO web pages for some time. I am sorry but I do not believe in big conspiracy theories and that the government is trying to hide something from us concerning Mars or any other planet. I believe that we still have a lot to learn from Mars and that we should keep our minds as open as possible for interpretations that do not necessarily conform with what we are used to on Earth...

I do not regard the above as being an explanation – it does not make any arguments based on points of data, or science, so was rather disappointing in this regard. The geologist said, at the end of his message:

Good luck with your investigations!

I took this to mean that he did not wish to investigate this strange geological (or archaeological) formation himself, so any further dialogue would have been pointless.

A friend suggested that the dome may have formed from some kind of lava tube. Clearly, this is not an explanation that is based on a rigorous analysis of the available data – but then again, my friend is not a professional scientist (and neither am I). However, it is at least an idea - a starting point on which a model could be built – possibly considering that vulcanism on Mars is, according to existing models, much different than that at work on Earth.

I had 2 further responses from the 42 geologists and one expressed interest in what appeared to him, initially at least, to be mysterious, but he never replied to offer me an explanation. This, of course, exactly parallels the experience of Alan Moen, regarding his analysis of evidence of trees on Mars, that we discussed in chapter 12.

Someone that Knew Me

Here is another example of how the consensus view influences and controls behaviour and thinking. Again, back in 2004, I sent some of the same Mars images I mentioned in the last section to a fellow Open University tutor. At the time, this tutor was studying Radar Altimetry. (This type of instrument has been used on some of the orbiters, although MGS had a laser altimeter rather than a radar altimeter.) I was trying to arouse her curiosity in the anomalies shown in the images. I only included links to NASA or USGS websites. Her response to my message about what the images were showing was quite interesting.:

> *"It could have got into that section by accident or mischief"*

I asked, "Would you not think it irresponsible of NASA to shove in a few 'faked images' in their own data?" The response came:

> *"yes, but it's just what I could imagine a bored tech doing for 'fun' - even if irresponsible"*

I remarked, "Of course, it's a possibility, but I think if news of this leaked out, it would be quite significant - and a blow to NASA's credibility. How could we then know which images were the 'fun' ones and which were the 'real' ones? What yardstick would we use to judge them?"

I got no further suggestions in response to this comment.

An Overt Example of the Control of Information?

In the 1970s and 1980s, to obtain good quality copies of space-related images, NASA or similar organisations had to be contacted directly and then reprints or photocopies would be sent to those requesting them.

In the mid to late 1990s, the internet became much more widely used and so it was expected that NASA, and similar bodies, would post images on websites. By then, Richard Hoagland, Dr Mark Carlotto, Dr Brian O'Leary and others had already begun to seriously challenge NASA using their own data. It is interesting, then, that it was about the same time that NASA gave control of all the Mars imagery to Michael Malin and his private company, Malin Space Science Systems. Malin was then permitted to withhold images until he wanted to release them. Was this step implemented because when the new high-resolution Mars Orbiter images arrived on earth, unlike many of the earlier missions, these images would be downloaded and studied more quickly by thousands of people?

It is curious, then, to note that the rover images were still released directly through JPL/NASA. Can I be so "paranoid" as to suggest that this is because the rovers aren't actually on Mars and someone in NASA knows exactly where they are, and what they are likely to find? Whatever they have found (evidence of liquid water, fossils etc) has been easily controlled by experts shrugging their shoulders and simply ignoring any evidence of biology or palaeontology.

NASA Visits My Website

If you have studied the evidence in this book, it should be fairly clear by now that there is a story we are not being told. NASA's "flip-flop" behaviour regarding the discovery/un-discovery of liquid water, methane and evidence of microbial life on Mars amounts to gate-keeping.

Similarly, if you read Dr Mark Carlotto's "The Cydonia Controversy" book, you will see the ongoing pattern of denial, ignorance of evidence and reluctance to do the sorts of studies that people outside NASA have been undertaking for decades.

NASA seems to ignore the most interesting data and images it obtains and cling to the standard chronology and model of the formation of the Solar System and the presence of (supposedly) intelligent life within it.

As you will see below, they *do visit* websites like mine from time to time, which I find quite amusing. They are clearly hiding something – and it might only be a desire to keep their funding model in operation, which is quite similar to how, for example, cancer research operates. That is, their undisclosed basis of operation is one of "self-preservation," which comes before "finding and telling the truth."

Domain Name	nasa.gov ? (U.S. Government)		
IP Address	163.205.156.# (National Aeronautics and Space Association)		
ISP	National Aeronautics and Space Association		
Location	Continent	:	North America
	Country	:	United States ▧ (Facts)
	State	:	Florida
	City	:	Cape Canaveral
	Lat/Long	:	28.3898, -80.6051 (Map)
Time of Visit	May 15 2008 9:26:58 pm		
Last Page View	May 15 2008 9:32:33 pm		
Visit Length	5 minutes 35 seconds		
Page Views	1		
Referring URL	http://search.aol.co...-3502038853372907502		
Search Engine	search.aol.com		
Search Words	news life on mars		
Visit Entry Page	http://www.checkthee...tent&task=view&id=33		
Visit Exit Page	http://www.checkthee...tent&task=view&id=33		
Out Click	http://www.checktheevidence.com/audio/Rush-Limbaugh-March%204-2004-The%20Mars-Gore-Report.mp3		
	http://www.checkthee...Mars-Gore-Report.mp3		
Time Zone	UTC-5:00		
Visitor's Time	May 15 2008 4:26:58 pm		

Domain Name	nasa.gov [2] (U.S. Government)
IP Address	128.158.1.# (National Aeronautics and Space Association)
ISP	National Aeronautics and Space Association
Location	Continent : North America Country : United States ▦ (Facts) State : **Alabama** **This is the Marshall** City : **Huntsville** **Space Flight Centre** Lat/Long : 34.7018, -86.6108 (Map) Distance : 4,165 miles
Operating System	Microsoft WinXP
Monitor	Resolution : 1152 x 864 Color Depth : 32 bits
Time of Visit	Jan 16 2009 2:33:14 pm
Last Page View	Jan 16 2009 2:45:44 pm
Visit Length	12 minutes 30 seconds
Page Views	1
Referring URL	
Visit Entry Page	http://www.checkthee...view&id=33&Itemid=59
Visit Exit Page	http://www.checkthee...view&id=33&Itemid=59
Out Click	http://www.msss.com/moc_gallery/ab1_m04/images/SP243004.html http://www.msss.com/...images/SP243004.html
Time Zone	UTC-6:00
Visitor's Time	Jan 16 2009 8:33:14 am
Visit Number	119,779

NASA Visits Checktheevidence.com – the second visit led them to their own "T"-shaped Mars anomaly seen on image SP243004.

"NASA and ESA Wouldn't Lie to the Public"

In a number of places in the Hoagland/Bara book "Dark Mission, " a 1959/1960 report by the Brookings Institution called "Proposed Studies on the Implications of Peaceful Space Activities for Human Affairs"[507] is referenced, in relation to what would happen if a group like NASA actually discovered evidence of extraterrestrial life. (Note: there is also a lengthy collection of Footnotes to the main "Brookings" report.[508]) In "Dark Mission," it is rightly mentioned that the whole report is over 200 pages long, and considers many issues, not just the discovery of extraterrestrial life. It does contain some interesting quotes, which I will describe below. I do find it interesting that the concept of extra-terrestrial life was mentioned at all, though. For example, on page 2S, in point 5 we can read:

> *Certain potential products or consequences of space activities imply such a high degree of change in world conditions that it would be unprofitable within the purview of this report to propose research on them. Examples include a controlled thermonuclear fusion rocket power source and face-to-face meetings with extraterrestrials.*

The report even references the sort of thing we have discussed in chapters of this book. On page 42S, Point 4:

> *Though intelligent or semi-intelligent life conceivably exists elsewhere in our Solar System, if intelligent extraterrestrial life is discovered in the next twenty years, it will very probably be by radio telescope from other Solar Systems. Evidences of its existence might also be found in artifacts left on the moon or other planets.*

The report considers how cultures would react to the discovery of extraterrestrial life:

> *The consequences for attitudes and values are unpredictable, but would vary profoundly in different cultures and between groups within complex societies; a crucial factor would be the nature of the communication between us and the other beings. Whether or not earth would be inspired to an all-out space effort by such a discovery is moot: societies sure of their own place in the universe have disintegrated when confronted by a superior society, and others have survived even though changed. Clearly, the better we can come to understand the factors involved in responding to such crises the better prepared we may be.*

These considerations continue on page 44S:

> *While the discovery of intelligent life in other parts of the universe is not likely in the immediate future, it could nevertheless happen at any time. Whenever it does occur its consequences for earth attitudes and values may be profound. Hence a long-term research effort, which would aid in preparing for this possibility, could usefully begin with:*
>
> > *A continuing determination of emotional and intellectual understanding and attitudes regarding the possibility and consequences of discovering intelligent extraterrestrial life...*

Much later in the report on page 183, we read:

> *Since intelligent life might be discovered at any time via the radio telescope research presently under way, and since the consequences of such a discovery are presently unpredictable because of our limited knowledge of behavior under even an approximation of such dramatic circumstances, two research areas can be recommended:*
>
> > • *Continuing studies to determine emotional and intellectual understanding and attitudes -- and successive alterations of them if any -- regarding the possibility and consequences of discovering intelligent extraterrestrial life.*
>
> > • *Historical and empirical studies of the behavior of peoples and their leaders when confronted with dramatic and unfamiliar events or social pressures. Such studies might help to provide programs for meeting and adjusting to the implications of such a discovery.*
>
> *Questions one might wish to answer by such studies would include: **How might such information, under what circumstances, be presented to or withheld from the public for what ends? What might be the role of the discovering***

scientists and other decision makers regarding release of the fact of discovery?

Richard Hoagland makes quite a bit of this last quote which discusses the idea of withholding information from the public. We should note, however, that the report does not overtly recommend that this information should be withheld, it simply lists it as a consideration. I don't think we can use this to say that this Brookings Institution report suggests that discoveries proving the reality of extraterrestrial life should be covered up (even though this, from where I am standing, appears to be exactly what has happened). To me, the report suggests a policy not dissimilar from how you might report the news of a sick or dying relative or friend to a loved one – you might not want to upset the person, so you might not tell them the whole truth.

Taking the view that "Brookings" *is* recommending the cover up of the discovery of extra-terrestrial life, it can easily be argued by sceptics that these quotes are from an "old-fashioned" point of view. That is, the report was written at the dawn of the space age – even before the first man had gone into orbit. According to the official accounts that apologists have to support, then, the story is that "we have been into space and found no evidence of extraterrestrial life – present or past, so the Brookings institution report is simply out of date." In essence, then, we end up with a circular argument – i.e. the Brookings report was written before we "knew there was no other life in the Solar System" and "we have found no evidence – so any recommendations it makes aren't relevant now."

Perception Management

Another well-known event that is mentioned in the Hoagland/Bara "Dark Mission" book is the reaction to the Orson Wells 1938 Radio Play "War of the Worlds."

> *The [Brookings] Report then references an obscure work by psychologist Hadley Cantrell, titled "The Invasion From Mars: A Study in the Psychology of Panic (Princeton University Press, 1940)." The Rockefeller Foundation under a grant to Princeton University commissioned this little known book. Its subject was the 1938 Orson Welles War of the Worlds broadcast (**which it is estimated that more than a million people in the northeast United States panicked over**). The implication is that the broadcast was a warfare psychology experiment, and that America dramatically failed the test.*

Hoagland's spelling has gone slightly adrift here – as the author of the study is "Cantril." Hoagland is referring to footnote 37, included in the "page 183" quote I referenced above. Footnote 37 reads as follows:

> *37. Such studies would include historical reactions to hoaxes, psychic manifestations, unidentified flying objects, etc. Hadley Cantril's study, Invasion from Mars (Princeton University Press, 1940), would provide a useful if limited guide in this area. Fruitful understanding might be gained from a comparative study of factors affecting the responses of primitive societies to exposure to*

technologically advanced societies. Some thrived, some endured, and some died.

What interested me about this is that recent research by Carl James has referenced this very same report, and what he has found out is that there was probably no mass-panic at the Orson Welles radio play. Only a few people "ran into the streets." This research is based on letters sent to radio networks and other media outlets following the 1938 Orson Wells broadcast. [509]

I might suggest, then, that the idea of "mass panic" being caused by the discovery of extra-terrestrial life has been deliberately over-stated. The effect of the Cantril report *may* have made it *more likely* that a policy of covering up discoveries and evidence, such as what we have covered in this book, would be adopted. Said another way, they used the public reaction to the "War of the Worlds" radio play as an excuse for keeping secrets. (However, it is a complicated issue, to say the least!)

"Denial is not just a river in Egypt…"

Reading about the experiences of Dr Mark Carlotto, Dr Brian O'Leary, Richard Hoagland and others while they were undertaking their own Cydonia investigation is quite reminiscent of other research I have been involved in. It is research that, in some way, threatens the established institutions - of science or religion. The members of the academic community have a strong desire to deny or dismiss the evidence and when they do agree to analyse it, impossible standards of proof are set. This is very similar to what I observed with Lloyd Pye's research into Human Origins[510] and the so-called Starchild Skull[511]. Again, I encountered similar reactions to Dr Judy Wood's research into the destruction of the World Trade Centre on 9/11[512]. Trained minds often refuse to look at any of the evidence. When they do look, they will perhaps be satisfied if they can explain one or two elements of the evidence and then "consider the matter closed."

Divide and Conquer

In the "Dark Mission" book, Hoagland and Bara discuss the progress of their Cydonia research and how, for example, disagreements arose over the evidence and they began to distance themselves from each other. For example, Professor Stanley McDaniel, who invested so much time in completing the critique of NASA's handling of the Cydonia data and research, eventually split from the rest of the group and no longer seemed to support the research (at least, not as vociferously as he did when he first got involved). Similarly, there was a split between Hoagland and Van Flandern when they disagreed over the content of certain MGS images which became available in 2001. On page 312 of "Dark Mission," Hoagland discussed how he found lettering on the D & M Pyramid. Tom Van Flandern found the same letters. Hoagland was more or less convinced that someone at NASA or MSSS had doctored the images while Van

Flandern thought the letters were "really there," etched on the surface of the pyramid. Similarly, on page 332 of "Dark Mission," Hoagland discusses the "Nefertiti head" image which Van Flandern showed at the 2001 press conference. Other disagreements meant that Hoagland did not attend or speak at this conference. (I looked at the "Nefertiti" image and the "other face" which appears in a cliff, and I agree with Hoagland that this wasn't compelling evidence of artificiality and should not have been shown at the press conference without more detailed study, which had not been done.)

In chapter 11 of the "Dark Mission" book, Hoagland also goes on to describe how he "fell out" with Dr Mark Carlotto over disagreements regarding the processing of the 2002 THEMIS image of the Face, which I mentioned in chapter 4. Similarly, Holger Isenberg teamed up with Hoagland and Keith Laney in doing an analysis of the same THEMIS image, but then they later had irreconcilable differences over the same data and no longer worked together.

The factors outlined here, as well as many others, contribute to keeping the cover up in place.

23. Gatekeepers on Earth

Popular figures in Science in the USA and UK will never properly discuss the sort of evidence contained in this book. When the subject of extraterrestrial life comes up, the discussion is fairly predictable. For example, if the "Face" on Mars is brought up, a proclamation will be made that it "was found to be a normal mesa." If the anomalous SOHO images are brought up, it will be stated that they are the result of "comets or cosmic rays." It really doesn't matter what evidence you want to talk about, the "experts" always know better. How could they not know better? It is these "experts" that are invited to discuss things on "tee vee," so they must know the truth of things, mustn't they?

Dr Steven Hawking

Hawking has been a prominent cosmologist since the 1970s – this is a bit of a mystery, in of itself, as he has outlived all those who have the same illness by an order of magnitude in years. This is so unusual that it's been commented on in mainstream media articles. For example, an article in the Washington Post from 24 Feb 2015 states[513]:

> And then there's Hawking. He has passed that two-decade mark twice — first in 1983, then in 2003. It's now 2015. His capacity for survival is so great some experts say he can't possibly suffer from ALS given the ease with which the disease traditionally dispatches victims. And others say they've simply never seen anyone like Hawking.

Hawking has now been known to comment about alien life, such as was reported in an article in the UK Guardian on 30 April 2010[514]:

> "If aliens visit us, the outcome would be much as when Columbus landed in America, which didn't turn out well for the Native Americans," Hawking has said in a forthcoming documentary made for the Discovery Channel. He argues that, instead of trying to find and communicate with life in the cosmos, humans would be better off doing everything they can to avoid contact.

This, of course, echoes what was said in the 1960 Brookings Institution report that was covered in chapter 22. It is just an opinion and is not based on any evidence. Hawking doesn't reference any of the research done into the issue – such as what is in this book (evidence which has primarily come from NASA). Neither does he reference things like the COMETA report[515] or any other relevant literature.

Perhaps it is inappropriate or even cruel of me to suggest that someone who is immobile and cannot speak without the aid of a speech synthesiser could easily have "words put into his mouth." There is information available online which explores this concept in great detail.[516]

Dr Brian Cox

Cox is a ubiquitous figure on British television and is regularly quoted in the press. I will leave readers to do their own research on this fellow, but for now I will just include a quote from an article from the Wired Magazine Website from 29 November 2017[517].

> On the likelihood of detecting life beyond Earth
>
> *I think the chances of detecting microbial life beyond Earth are high. If we went to Europa and went to Enceladus and went to Mars and had a good look, I wouldn't be surprised if in one or more of those places you find microbes.*

Clearly Cox is either ignorant of the experiments already sent to Mars (see chapters 10, 11 and 12) or he has chosen not to talk about them. How would we ever get someone like Brian Cox to tell the truth? Perhaps someone will take him on a long car journey and force him to read this book...[518]

Cox also parroted mainstream opinions and indulged in evidence-denial, in an earlier article in the Sunday times on 9 October 2016[519]:

> *Mankind's search for alien civilisations may never succeed, because intelligent life destroys itself not long after it evolves, Professor Brian Cox has suggested.*
>
> *He was addressing one of astronomy's great mysteries: why, given the estimated 200bn-400bn stars and at least 100bn planets in our galaxy, are there no signs of alien intelligence?*

Again, this is opinion – based on a comprehensive and thorough denial of available evidence.

Dr Carl Sagan

I first heard of Sagan when his ground-breaking "Cosmos" series was aired on UK TV in the early 1980s. However, he had been involved since the 1960s in discussions about UFOs, aliens and extra-terrestrial life.

Of course, we have already mentioned Dr Sagan many times in earlier chapters and if you read all the quotes, a strange picture emerges. He expressed a number of different viewpoints over the years – and seemed to, perhaps, vary his stance depending on what audience he was addressing[520]. Let us not forget, for example, that he even wrote a novel called "Contact" (first published in 1985)[521] about humans actually *making* contact with an alien civilisation... This novel was made into an interesting film/movie in 1997[522] (just after Sagan's death).

Here are a couple of Sagan's quotes, included in Dr Mark Carlotto's "Cydonia Controversy" book:

> *"The question is not whether you are right or wrong, sir. You are not even in the conversation."*
>
> *Dr Carl Sagan to Dr John Brandenburg regarding Brandenburg's work on Cydonia*[523]

Famously, Carl Sagan said, denouncing the claims of potential artificiality on Cydonia:

> *Extraordinary claims required extraordinary evidence.*

What is "extra-ordinary evidence"? A large amount of evidence, perhaps? Extremely powerful evidence, maybe? The problem is these definitions like the parent phases are ambiguous and relative. Sagan, as an articulate speaker should have been aware of this.

Of particular interest to me is Sagan's involvement in a debate over a proposed UFO Symposium to be hosted by The American Association for the Advancement of science[524]. This was an event where UFO researchers wanted to present evidence before a panel of Scientists, following the strong dissatisfaction expressed by many over the Condon Report.[525] There was significant opposition to the Symposium being held at all, and it took almost a year to get the AAAS and others to agree to be involved. Sagan even introduced a presentation by Dr James MacDonald that was highly critical of earlier government reports (Blue Book, Project Sign, Project Grudge etc) on the UFO issue.[526]

Another person that was involved in the 1969 AAAS symposium was astronomer Donald Menzel. He was a hardened UFO debunker. A report about the organising of the AAAS 1969 symposium[524] summarises Menzel's actions and notes that (a) he drafted 1968 protest letter about the meeting, (b) said he "would not appear with crackpots," (c) suggested cancellation of the symposium if a one-sided program were not possible and (d) he asked for speaking time equivalent to two speakers who were advocating the validity and importance of scientific research into the UFO phenomenon.

I bring this up because it has been reported by UFO researcher Stanton Freidman that he found some evidence that Menzel was a member of MJ-12[527] – a secret group that was set up to manage the cover up that the US government and military undertook, following the events in Roswell, New Mexico in July 1947[528].

It has been suggested that Sagan may have got similarly involved[529] due to him doing secret work on a classified project to drop a nuclear bomb on the moon[530]! This means he would have had a top-secret security clearance, so he wasn't just a "TV astronomer."

TR-59-39

Vol I

(Unclassified Title)

A STUDY OF LUNAR RESEARCH FLIGHTS

by

L. Reiffel

With contributions by

R. W. Benson	J. J. Brophy
I. Filosofo	N. S. Kapany
D. Langford	W. E. Loewe
D. Mergerian	O. H. Olson
V. J. Raelson	C. E. Sagan
P. N. Slater	

19 June 1959

Research Directorate
AIR FORCE SPECIAL WEAPONS CENTER
Air Research and Development Command
Kirtland Air Force Base
New Mexico

Approved·

Fred A. Gross, Jr.
for DAVID R. JONES
Lt Colonel USAF
Chief, Physics Division

Leonard A. Eddy
LEONARD A. EDDY
Colonel USAF
Director, Research Directorate

Project Number 5776-57853

Contract AF 29(601)-1164

1959 document with Sagan's name - from the Air Force Special Weapons Center - Air Research and Development Command Kirtland Air Force base. New Mexico

Perhaps this, then, explains why he had an important gate-keeping role in relation to the discovery of evidence of extraterrestrial life – he had previously had a security clearance and had done work for the US military. It is interesting to note, however, that Sagan was a critic of the spending on nuclear weapons and the SDI programme.[531] That said, it was easy to publicly be "anti-nuclear warfare" – in the 1980s – and now. Sagan who was a consummate communicator, would have been good at doing "muddle ups," by appearing to be sympathetic to both sides of the arguments, but never firmly supporting the most compelling evidence that became available.

Neil deGrasse Tyson

He is another popular science speaker and author. From his website:

> Neil deGrasse Tyson was born and raised in New York City where he was educated in the public schools clear through his graduation from the Bronx High School of Science. Tyson went on to earn his BA in Physics from Harvard and his PhD in Astrophysics from Columbia.
>
> Tyson's professional research interests are broad, but include star formation, exploding stars, dwarf galaxies, and the structure of our Milky Way.

Again, like Dr Brian Cox in the UK, he doesn't seem to be aware of the evidence gathered by NASA since 1976. He talks about the possibility of water on Mars in the past and the implications of detecting life. He does not talk about any of (NASA/JPL/ESA) evidence that shows life has already been found on Mars. For example, in a UK Guardian Newspaper article, dated 30 October 2016, he states[532]:

> I'd say we need to know a little bit more about nearby planets, like Mars, and some nearby moons within our Solar System, like Europa around Jupiter or [Saturn's] Enceladus. These have tantalising properties that could support life as we know it, but in its extremes – extreme cold, extreme dehydration or extreme radiation. And if we find life in our Solar System, that bodes very well for life being ubiquitous.

In an undated posting on the American Natural History Museum's website[533] (but likely posted in late 2012, as the Internet Archive's oldest capture is 02 January 2013[534]) deGrasse Tyson is quoted as saying:

> **What have Spirit and Opportunity, NASA's two Mars Exploration Rovers, been discovering up there?**
>
> Until now, all the evidence for water on Mars has been strongly circumstantial. It looks like water was there, because the surface features resemble features on Earth that we know are made by water. ... the most recent evidence from the rovers confirms what we strongly thought before, that in fact water enjoyed a major presence on Mars.... The current rovers aren't carrying biological experiments; they're carrying chemical experiments. One of the chemical experiments involves identifying certain kinds of rocks that can form only in the presence of water. Those rocks were found-so that's an important first step.
>
> **Suppose fossils or other signs of life are discovered on Mars. What are the implications?**
>
> There is evidence to suggest that Mars was wet before Earth was wet. If that's the case, maybe Mars had life before Earth had life. ... you have to admit the possibility that life traveled from Mars to Earth as stowaways in the cracks of rocks. Which would make all humans — all life on Earth — the descendants of Martians.

What do you think are the odds of finding evidence of life on Mars?

High — certainly higher than 50 percent. And by life, I refer to fossil life, not current life; and simple life, bacteria or microbes. Water is the key.

So, again, he speculates and doesn't give a clear picture of the evidence already gathered – and referenced in this book (and elsewhere!) He makes no mention of the Viking, ESA and Curiosity experimental results. He should know about these – and be able to discuss them authoritatively and accurately.

Richard C Hoagland

Perhaps a surprising name to be on this list, due to the number of times I have quoted him in this volume. Clearly, he has done an enormous amount of work exposing Mars and Solar System artefacts and anomalies. Despite this, Hoagland's name has to be added – due to his promotion and belief in the Apollo hoax. Hoagland is an example of a researcher that has closely studied NASA imagery and exposed some of the lies they have told. Why is it, then, that he continues to promote the Apollo landings as having really happened, when they cannot have happened as stated? Obviously, I mentioned the hoax in chapter 17 and I will likely cover this in detail in a separate book, but readers can work through my presentation called Apollo: Removing Truth's Protective Layers[26] for further information if they wish.)

In chapter twelve of his "Dark Mission" book, Hoagland even shows examples of faked Apollo images (which were published in Alan Shepard's book "Moon Shot.") This means that Hoagland does realise NASA has faked some images. His promotion of the rest of the Apollo programme as being real is then, to me, rather less understandable, bearing in mind how much he has studied many of these photos.

Further, Hoagland promotes many claims which go considerably beyond what can be supported by the evidence. One only needs to read the introduction to his book "Dark Mission" to understand this. He, often, "rants" about his findings and in my opinion, his verbose and often "fuzzy" presentations dilute the strength of many of the points he makes.

By making exaggerated and unsupported claims, and false claims (such as those in Chapter 5 of his "Dark Mission" book[535]), he makes it much easier for debunkers to be used to limit or silence discussion of the very anomalies he is trying to draw attention to. Similarly, the good/truthful information he presents (such as the information about occult and Masonic practices within NASA) gets mixed in with the bad. It is then far less likely that this evidence undergoes serious consideration, by those who have some scientific training or education, but consider stories about Masonic conspiracies or Solar System anomalies are just "silly." In other words, it seems to me that Hoagland is just "controlled opposition" to those who want to present only the most compelling evidence that we have not been told the truth about discoveries in the Solar System.

This is also illustrated in his actions, in 2011, regarding Dr Judy Wood's 9/11 research[536], which I wrote about on my website[95], and in my free eBook, "9/11 Finding the Truth[96]". In his 2011 presentation at a conference in the Netherlands, Hoagland expresses scepticism of the official story, yet in chapter 12 of his "Dark Mission" book, he essentially parrots the official narrative and links this in with his discussion of George W. Bush's "rebooted" space programme. This sort of problem is not limited to Richard Hoagland, though – and is yet another factor which affects how "gates are kept closed."

One must also question the whole basis from which his Cydonia investigation started. On page 472 of his book "The Monuments of Mars: A City on the Edge of Forever," Hoagland mentions that it was SRI who funded some of his research. He also mentioned this in a 1989 CompuServe Chat[537],

> *Then, in 1983 I got the images, looked at them and realized that I was seeing not only a "face" but other things which didn't belong there. That's when the third IMAGING investigation began -- at SRI. Their Prime system did some excellent work, using digital copies of the NASA Viking tapes. That's when we turned to a FOURTH source, for "independent data" on the imaging: Mark Carlotto. Mars Inv team done under auspices of SRI*

Further research reveals that it was probably Dr Lambert Dolphin, who worked at SRI, who set this up[538]:

> *Hoagland's interest in Giza began when he met Dr Lambert Dolphin, a scientist from the Stanford Research Institute (SRI) who had conducted radar and seismographic research to locate hidden chambers near and under the Great Sphinx.*
>
> *Dr Dolphin was captivated by Hoagland's idea of a 'Martian Sphinx' - the Face - and , in 1983, was instrumental in helping Hoagland set up the Independent Mars Investigation Project under the aegis of SRI.*

Dolphin has an interesting resume, where he states he worked for SRI for 30 years.[539]

SRI is significant, I believe, because a number of people whom I think have much more knowledge than we do about "fringe issues" have worked for them. For example, it was at SRI that the remote viewing program which became known as "Operation Stargate" had its genesis[540]. SRI has also investigated Cold Fusion[541]. Another two key figures that have been connected, at one time or another, to SRI that are worthy of further investigation are Col John B Alexander[542] and Harold Puthoff[543].

My guess is that Hoagland was used so that people within SRI, or those "in the shadows" could "keep an eye" on what he was doing and possibly "steer" or influence the progress of his Mars Research.

Hoagland has repeatedly said "The lie is different at every level." How ironic it is that Hoagland, himself, will not tell the whole truth about the manned and unmanned space programme.

Arthur C. Clarke

In chapter 12, we noted Clarke's stance on the possible "banyan tree" MGS images from 2001. Other researchers have written about Arthur C. Clarke's possible knowledge of a "secret history" of our Solar System. Indeed, Richard C Hoagland's "Moon With A View Series" about Iapetus[544] is subtitled "Or, What Did Arthur Know ... and When Did He Know it?" If one studies the film/movie "2001: A Space Odyssey," it seems there could well be layers of hidden meaning, as researchers like Jay Weidner have suggested[545]. Did Arthur C. Clarke, as author of "The Sentinel," on which the story of the film was based, know more than he let on? In Clarke's "Sentinel" story, Wilson, a geologist exploring the Moon, tells of the experience of discovering a pyramid shaped object that indicates the presence of extraterrestrial intelligence. He describes the experience of discovery and creates a story that reflects on the meaning of the object.[546] Did Clarke also have knowledge about Iapetus, well before the discoveries made by the Cassini Probe in 2004?

On 10 September 2007 (only a few months before his death), Arthur C. Clarke made a 5-minute "Video greeting" to NASA/JPL to mark the closest Iapetus flyby of the Cassini spacecraft.[547] He read a piece to camera, which included the following words.

> *Thanks to the World Wide Web, I have been following the progress of Cassini-Huygens mission from the time it was launched several years ago. As you know, I have more than a passing interest in Saturn.*
>
> *And I was really spooked in early 2005, when the Huygens probe returned sound recordings from the surface of Titan. This is exactly what I had described in my 1975 novel Imperial Earth, where my character is listening to the winds blowing over the desert plains.*
>
> *[the 2007 flyby] is a particularly exciting moment for fans of "2001: A Space Odyssey" - because that's where the lone astronaut Dave Bowman discovers the Saturn monolith, which turns out to be a gateway to the stars... More than 40 years later, I cannot remember why I placed the Saturn monolith on Iapetus. At that time, in the early days of the Space Age, earth-based telescopes couldn't show any details of this celestial body. But I have always had a strange fascination for Saturn and its family of Moons.*
>
> *But in the movie, Stanley Kubrick decided to place all the actions at Jupiter, not Saturn. Why this change? Well, for one thing it made a more straightforward storyline. And more important, the special effects department couldn't produce a Saturn that Stanley found convincing...*

Clarke ends the greeting by saying:

> *...And who knows, one day our survival on Earth might depend on what we discover out there.*

Gatekeeping... or Disclosure?

In chapter 14, we read about the strange foreknowledge about the moons of Mars, which was apparently possessed by Jonathan Swift, author of "Gulliver's Travels." In this section, we will look at two examples of Science Fiction Artwork and a couple of examples of Comic Art which appear to show a similarly strange foreknowledge of certain things in the Solar System.

A Comic Face

In 1958, "The Face on Mars" story was published in a comic titled "Race for the Moon." The story was by Jack Kirby and inked by Al Williamson. From a comic data website, we read[548]:

"On an expedition from the Earth's moon to the planet Mars, an international team of astronauts – led by American Ben Fisher – discover a huge carving of a Martian face – that's as big as a mountain! ... Fisher explains that the statue contains "a visual history of a race's heroic death – and the triumph of a surviving memory". Later, as they pilot their rocket to Jupiter, to Earth, Fisher and his team take careful notice of the debris-strewn asteroid belt – "the pieces of a planet that blew up between Mars and Jupiter".

Wayne Herschel has written about this 1958 story on his website[549]. Herschel writes about a "forehead anomaly" on the face, which is also, allegedly, shown in the comic's image of the face (see above). However, I am not sure I can see this anomaly in the MGS facial image, or other images of the Face.

A Second Science Fiction Face

In 1953, an Arthur C Clarke Science Fiction Story called "Against the Fall of Night" (also known as The City and the Stars)[550] was published. It was originally a shorter story, which was expanded to a full novel. There were various re-printings of this, but the 1970 version is of interest, due to the cover[551]. Here we see another apparent example of "premonition".

Though this story is not set on Mars, it does seem rather peculiar to see a sculpted face, of enormous size, laying near what might be a city (or at least a few buildings) with the rising sun over the brow of the nose of the face. This version came out in August 1970 and the illustrator was Ron Walotsky, though he was not credited.

Did Ron Walotsky "tune in" to something, or was someone giving him ideas…? We will probably never know…

Smiley Face Crater

A blog by Scott Jeffery from 2012[552] notes another odd "coincidence" relating to the mid-1980s "Watchmen" comic strip, created by Alan Moore and Dave Gibbons. One of the creators is quoted thus:

"we didn't know that there was a smiley face on Mars. We discovered halfway through that there was a crater on Mars that looked like a giant smiley face".

Below, I have included an image from Google Mars, showing the "Smiley Face," at lat 50.97°S, lon 31.77°W.

Note how several features seem to match quite well – such as the "fuzzy" right eye and similarly sized crater to the bottom left of the "smiley face" one.

The crater just below the "smiley face" appears to have a hexagonal appearance and there is also a linear ridge just to the south-west of that crater.

Scott Jeffery's article discusses other similar instances of premonitions in Comic Art.

Iapetus and… George Lucas

Richard C Hoagland was, again, the first to note[553] the similarity shown below.

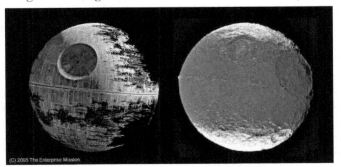

The "Death Star" (left) as seen in Part VI of the original Star Wars films.

Others have, for example, pointed out that Saturn's moon Mimas has a similar large crater feature, but it lacks the equatorial ridge feature.

Left: A Cassini image (PIA12739) of Saturn's smallish moon Mimas[554] – with a large crater feature. (Several moons seem to have single large craters like this – perhaps you can find them…)

I hope you will agree that it's difficult to put down such a close correspondence in the features seen in this chapter just to pure coincidence. It appears some people can access – or are given - knowledge that is hidden from others.

Hollywood's Version of What Happened to Mars…

If readers study Carl James' "Science Fiction and the Hidden Global Agenda[555]," they will learn about the influence that the US Military and CIA have had in determining the content and/or script of many Hollywood films/movies. The list of films/movies that have been "looked at" by the CIA is longer than you might expect.

Of course, there have been many films over the last few years which have featured Mars in the title and/or storyline. We have 1950s B movies like "Invaders from Mars[556]," two versions of H.G. Wells' "War of the Worlds" and films such as "Total Recall[557]" (1990) and "John Carter of Mars[558]" (2012) and "The Martian" (2015) – which we mentioned in chapter 9.

More intriguing, however is the ending of the Touchstone/Disney film "Mission to Mars," released in 2000[559], directed by Brian De Palma, brother of the N-Machine's inventor Bruce De Palma[560]. The first three quarters of the film is a run-of-the mill space rescue mission story, featuring elements of problem solving / resolving and heroism. However, in the final 40 minutes or so of the film, things "liven up" a bit. Are the writers/directors playing with the ideas presented by Hoagland, O'Leary, Carlotto and all the others? Or are they actually telling us something more. I would argue that elements of the storyline which then unfold show that, at least, someone has been listening very carefully to what these people have said and shown. On Richard C Hoagland's website, he reports that a trailer for the films says[561]:

> "For 25 years, the government has concealed evidence of a life-like formation on Mars…"

However, I could not find this version of the trailer. Another version of the trailer[562] does not have these words.

So, if you have not seen this film, and want to watch it, I must now alert you to the "spoilers" below… so skip to the next section if you want to watch the film for yourself. (All pictures in this section are © Disney Corporation.)

When the rescue mission reaches the only survivor from the previous mission, who has been stranded there for months or years, they find he has been receiving strange transmissions. He realises they have been emanating from the Face! He explains that it wasn't a ruin, as such, it was just covered with material to disguise its true appearance.

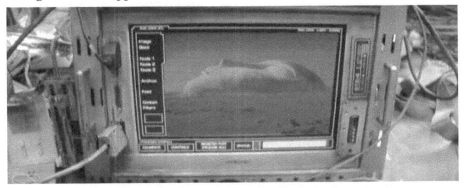

Following this, the 3 crew from the rescue mission and the survivor travel to the Face and send a coded message to get inside.

Once inside, they walk over a spectacular 3D/holographic Solar System simulation or recording, showing the planets orbiting the Sun. They see Mars as a blue planet, with oceans, but then, as they are walking past it, in the simulation, a fast-moving object – such as a comet or large asteroid – impacts into the surface.

A wave of destruction sweeps across the entire surface, turning the planet from blue to red.

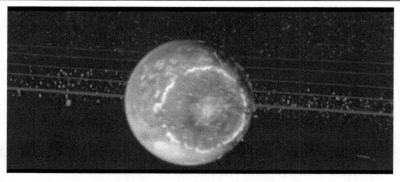

Following this, an 8-foot tall (feminine?) extra-terrestrial or ethereal being, golden in colour, approaches them from behind. They turn to see the being weeping. Moments later, the being reaches into its own chest and displays its right hand before them. A moving helical spiral of DNA sits in its hand.

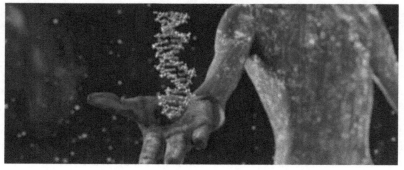

As this is happening, many golden sunflower seed-shaped spacecraft emerge from the planet and are seen to fly to a distant galaxy, but one ship stays behind.

The being then lifts its hand up to the single hovering ship and the DNA is transferred into the ship.

The ship then flies rapidly to the earth and immediately ditches in the ocean. All life on earth is then seen evolving from this event and one of the astronauts is heard to say....

"They seeded earth..."

Why did the "Mission to Mars" film end this way? After all, a note in the film's credits says:

> The National Aeronautics and Space Administration's cooperation and assistance does not reflect an endorsement of the contents of the film or the treatment of the characters depicted therein.

Might I suggest that NASA is, through the film, telling us a somewhat truthful story. It is very interesting, here, to consider what Disclosure Project Witness "AH" said in December 2000, as we covered in chapter 16. He also stated some additional things:

> They know for a fact that the face on Mars was made by an extraterrestrial race that came here to Earth about 45,000 years BC... and set up civilization here on earth and went back and forth from Mars to our planet Earth to give us information and to further prove the race that they created - which is us. This is this is a shock to the human populace. And I think it's the big reason why NASA and all the governments, especially United States, are refusing to release this information. Because one thing will lead to another and somebody will come to a conclusion that we were created by an extraterrestrial race that came to this solar system and set the planets in their proper order, which they are now in. And they exploded the gaseous planet that which, we call the Sun and thus "God said let there be light." This is this would be shocking to everybody here on earth and this is what they're afraid of releasing. ... I really feel that the aliens that were on planet Mars came here and set up our civilization ...

Was "AH" simply "telling a story" based on the plot of the "Mission to Mars" Film? Or was he disclosing some truthful information?

With the film, however, NASA automatically gains plausible deniability because the context in which they have put this information is a science fiction/adventure story. This works in 2 ways. Those like me, who have investigated the evidence for non-human beings interacting with us now and in the past, will see that the film contains a possible disclosure, but they don't know which parts are true. Uninformed viewers who come across evidence about, for example, the Face on Mars (as in the books by Carlotto et al) will subconsciously treat is as fiction/fantasy and therefore not study and take in the details. In both cases, the truth is not plainly revealed and so nothing really changes.

Outside of this, there are a number of versions of the story of the earth-Mars connection – and this comes almost exclusively from "channelled" sources – one such version can be found in an 11-minute video, posted in 2012[563] called "Atlantean Catastrophe". Of course, some people would laugh at this whole book, because I have even mentioned a video like this. However, people are searching for answers – and as I have shown, conclusively, in this volume, those that have been paid handsomely for their expertise, are not providing the most important answers. For whatever reason, they have repeatedly ignored evidence, ducked questions and even lied. This has resulted in an almost unassailable consensus and reductionist view of the Universe, which isolates our collective psyche and greatly inhibits our knowledge and understanding of how we came to be here and how we are connected into a much richer and more interesting picture of cosmic history.

24. Additional Thoughts, Conclusions

It is far better to grasp the universe as it really is than to persist in delusion, however satisfying and reassuring.

– Carl Sagan (1934-1996)

Before I describe my overall conclusions, I would like to cover, briefly, some topics that didn't seem to fit elsewhere in this volume.

Image Editing

When one starts to do the sort of "digging" that has been done in the creation of this book, one quickly begins to ask the question, "Does NASA or similar agencies modify images to cover things up?" The answer to this question is not straightforward. If we consider many of the images from the Hubble Space Telescope, for example, it is openly stated that these are modified, for good reasons. We can read about this in an article posted in 2010 on the Space.com website[564]:

> The observatory will often take photos of the same object through multiple filters. Scientists can then combine the images, assigning blue light to the data that came in through the blue filter, for example, red light to the data read through the red filter and green light to the green filter, to create a comprehensive color image.
>
> "We often use color as a tool, whether it is to enhance an object's detail or to visualize what ordinarily could never be seen by the human eye," NASA officials explain on the agency's Hubble Web site...
>
> "Creating color images out of the original black-and-white exposures is equal parts art and science," NASA said. For example, Hubble photographed the Cat's Eye Nebula through three narrow wavelengths of red light that correspond to radiation from hydrogen atoms, oxygen atoms, and nitrogen ions (nitrogen atoms with one electron removed). In that case, they assigned red, blue and green colors to the filters and combined them to highlight the subtle differences. In real life, those wavelengths of light would be hard to distinguish for humans.

So, some images have to be "constructed" for us to be able to see whatever is there. However, if those constructing the images found things they didn't want us to see, or that they thought were just "noise," would they remove them?

Taking an example from one of the Mars Rover Images (it's not clear whether it is from Spirit or Opportunity) did this contain a data drop out, is it some type of processing error, or has the image been "doctored" to remove something that someone didn't want us to see?

Left: Image file name
2P127520307EFF0309P2365L7M1 from
Spirit or Opportunity[565]

A further example of NASA Image editing was highlighted in an article by Peter Farquhar (Technology Editor) posted on 8 October 2010 on news.com.au. The article was entitled "Conspiracy theorists confident Photoshopped NASA image is a cover-up."[566] (We see use of another "gatekeeping" tactic creeping in here – with the use of the term "conspiracy theorists" even though the article explains how images *are* composited for effect. It subtly discourages investigation of images – as that sort of thing is only done by "crazy conspiracy theorists.") The article is about an image[567] posted earlier in 2010, on Astronomy Picture of the Day on 20 April 2010[568]. From the article, we see:

The NASA image of Saturn's moons Dione and Titan, left, and the high-contrast version showing Photoshop marks, right. Source:news.com.au

In the article we read (emphasis added):

*The person responsible for the manipulation, Emily Lakdawalla, told **a forum of excitable theorists** that she made the changes because of the way Cassini takes photos.*

> *"Cassini takes colour pictures by snapping three sequential photos through red, green, and blue filters," she said.*
>
> *"In the time that separated the three frames, Dione moved, so if I did a simple color composite I would be able to make Titan look right, but not Dione; or Dione look right, but not Titan.*
>
> *"So I aligned Dione, cut it out, and then aligned Titan, and then had to account for the missing bits of shadow where the bits of Dione had been in two of the three channels."*

Again, in this case, the explanation seems perfectly reasonable, but we have seen quite a few examples in this book where the reasons for changing, hiding or deleting the images are less clear.

Too Little Knowledge...? Too Much Knowledge...?

Several chapters in this volume have focused on the apparent ignorance of knowledgeable people when they are presented with evidence relating to the Cydonia Face, the Dome anomalies, evidence of life, fossils and so forth. Whilst they might accuse people like me of "seeing things which aren't really there" (pareidolia), there seems to be an "opposite" condition where people *cannot see* things that really are there. For me, this becomes clearest with things like the Mars dome that I covered in chapter 7), the fossil evidence in chapter 11 (wherever those fossils might actually be) and the features of Iapetus, described in chapter 19. It almost seems that if one has advanced degrees in science, one can become embedded in the specialist science itself and looking outside of this area of specialism becomes too difficult or impossible.

At the opposite and of the scale, however, I have also encountered liars – for example promoting the imminent arrival of "Planet X." Even when their predictions have failed and their lies have been demonstrated, they still persist with their lying. They have claimed to have taken photos of the Planet, but these photos show nothing. Added to that, Planet X cannot be seen in any terrestrial telescopes. Why can't people educate themselves to free themselves from these kinds of deceptions?

When one encounters attitudes and dishonest behaviour at each end of the scale, it becomes difficult to be optimistic about large numbers of people achieving any kind of enlightenment about what is *really* going on in the world – and outside of it.

Official Mars Discoveries – an incomplete Timeline

If we compile the list of discoveries and retractions into chronological order, we can see a pattern emerging.

Date	Discovery
1977[569]	Viking Labelled Release experiment detects microbial activity
1979[570]	Labelled Release Experiment results "revised"
18 May 1979[216]	Water Ice frost on Martian Surface
26 June 2000[571]	Liquid Water evidence
12 Jan 2004[220]	Mud/Water on Mars
25 Feb 2005[235]	Methane detected in atmosphere
6 Dec 2006[227]	Liquid Water Evidence
15 Jan 2009[237]	Methane detected
04 Aug 2011[228]	Liquid water flows
4 Dec 2012[238]	Curiosity Rover detects water and organic compounds on Mars
20 Sept 2013[239]	"No Life on Mars"
17 Dec 2014[240]	"Curiosity detects first burps of potential life on Mars"
29 Sept 2015[229]	Hints of Liquid Water

It is now over 40 years since the Viking Probe allegedly landed on the surface of Mars in 1976. Since then, several *billion* dollars have been spent on space exploration missions, yet if you read chapter 23, you will realise that what comes out of the mouths of astronomers and space scientists has barely changed. Are NASA and ESA simply in the business of self-preservation? If they downplay, ignore and even cover up the most important discoveries, what is the point of funding them? When a company has defrauded people out of millions or billions of pounds or dollars, and they get "found out," they are taken to court and may be "wound up" and the CEO's may even be thrown in prison (if they're not sufficiently well-connected into the groups which run planet earth, that is).

Due to the technical nature of much of the information in this volume and the fact that the majority of people have no interest in and don't care about space-related research (because they can't see how it affects or would affect their lives), the conmen get away with their massive, ongoing fraud.

Boredom

In presentations about Mars and "Secrets in the Solar System" that I have made, over the years, I have often included a summary I found on the "Above Top Secret" forum[572]. (This forum is itself largely co-opted and controlled by people in the shadows and is populated, mostly by anonymous posters.)

Let me give you the whole Mars conspiracy in a nutshell. I have no doubt that there is some real mind-blowing stuff going on on Mars and our governments know it. What they have to do is take the peoples' minds off of Mars. They send these rovers up there and they send back pictures of basically nothing. The TV and internet are flooded with these pictures of nothing. Soon the people get tired of looking at pictures of nothing and don't want to be bothered with Mars anymore. They trust that the government will make them aware of any important happenings on Mars should they occur. Of course, this will never happen.

We the taxpayers paid for those rovers. We had a right to know when they were launched and how they landed. We have a right to see the pictures, so they give us a million pictures of nothing. Now we're so bored with the pictures that we really don't care anymore. They know we won't ask questions anymore for fear of being bored out of our minds.

That is how it works. That is the trick to manipulating the people.

This summary, I would argue, applies not only to the Mars missions, but all of the space missions which have discovered evidence of present or past life in the Solar System. It also seems to apply to any discoveries which challenge the accepted version of how the Solar System was formed and what its history is. For example, even discussion of the exploded planet hypothesis does not seem to be encouraged in space science circles.

The Bottom Line

From the research I have been doing since 2004, I have made the following conclusions:

- We have not been told the truth about what has been discovered in our Solar System.

- The Solar System has a history – which likely includes activity on other planets and moons – by a past or current civilisation, which has some connection with humanity.

- Since the dawn of the Space Age, an ongoing cover up and perception management operation has been in progress.

- Disinformation has been published and circulated to confuse researchers and people wanting to know the truth.

- The group or groups operating this cover up would be considerably disadvantaged by any revelations – which are almost certainly related to the existence of advanced construction, propulsion and energy technologies.

- The status quo is primarily maintained using the services of the media, and the employment of gatekeepers, who will encourage and maintain a consensus which supports the secrecy and ridicules and/or attacks those who ask too many questions and/or make too many revelations.

There may be an additional reason for all the muddle-up and cover up that has been documented in this volume. It is my considered conclusion that a Secret Space Programme exists. I hope to cover the evidence for this conclusion in a future volume.

I hope you have found the information in this volume worthy of the investment of your time. I thank you for your attention and hope that you will share what you have learned with other knowledgeable friends, colleagues, associates and family members, whenever you feel comfortable in doing so.

References

In order to make access to links easier, I recommend you download an electronic version of this volume and access the links therein, or find a copy of the links on http://www.checktheevidence.com/ .

1 https://nssdc.gsfc.nasa.gov/nmc/spacecraftDisplay.do?id=1957-001B

2 https://history.nasa.gov/sputnik/

3 https://nssdc.gsfc.nasa.gov/nmc/spacecraftDisplay.do?id=1958-001A

4 https://nssdc.gsfc.nasa.gov/nmc/spacecraftDisplay.do?id=1959-012A

5 https://nssdc.gsfc.nasa.gov/nmc/spacecraftDisplay.do?id=1964-041A

6 https://nssdc.gsfc.nasa.gov/nmc/spacecraftDisplay.do?id=1966-006A

7 https://nssdc.gsfc.nasa.gov/nmc/spacecraftDisplay.do?id=1964-077A

8 http://www.enterprisemission.com/hoagland.html

9 https://news.nationalgeographic.com/2016/11/mars-exploration-nasa-mariner/

10 https://nssdc.gsfc.nasa.gov/imgcat/html/object_page/m04_10d.html

11 http://www.esa.int/Our_Activities/Space_Science/ExoMars/Schiaparelli_landing_investigation_completed

12 https://crgis.ndc.nasa.gov/historic/Lunar_Orbiter_Project

13 https://nssdc.gsfc.nasa.gov/nmc/spacecraftDisplay.do?id=1975-075C

14 https://nssdc.gsfc.nasa.gov/nmc/spacecraftDisplay.do?id=1995-065Ahttps://nssdc.gsfc.nasa.gov/nmc/spacecraftDisplay.do?id=1996-062A

15 https://nssdc.gsfc.nasa.gov/nmc/spacecraftDisplay.do?id=1996-062A

16 https://nssdc.gsfc.nasa.gov/nmc/spacecraftDisplay.do?id=2003-022A

17 https://nssdc.gsfc.nasa.gov/nmc/spacecraftDisplay.do?id=1997-061A

18 https://nssdc.gsfc.nasa.gov/nmc/spacecraftDisplay.do?id=2003-027A

19 https://nssdc.gsfc.nasa.gov/nmc/spacecraftDisplay.do?id=2007-034A

20 https://nssdc.gsfc.nasa.gov/nmc/spacecraftDisplay.do?id=2005-029A

21 https://nssdc.gsfc.nasa.gov/nmc/spacecraftDisplay.do?id=2009-031A

22 https://nssdc.gsfc.nasa.gov/nmc/spacecraftDisplay.do?id=2011-070A

23 https://www.amazon.co.uk/dp/B01MG7R0UO/ref=dp-kindle-redirect?_encoding=UTF8&btkr=1

24 https://www.youtube.com/watch?v=3KBYiWfHcLA

25 http://news.bbc.co.uk/1/hi/special_report/1998/03/98/gagarin/72184.stm

26 http://www.checktheevidence.co.uk/cms/index.php?option=com_content&task=view&id=324&Itemid=63

27 http://mcadams.posc.mu.edu/ike.htm

28 http://exopoliticshongkong.com/uploads/Ingo_Swann_-_Penetration_-_The_Question_of_Extraterrestrial_and_Human_Telepathy.pdf

[29] https://ntrs.nasa.gov/search.jsp?R=19680018720

[30] http://iopscience.iop.org/article/10.1088/0004-637X/697/1/1/meta

[31] https://crgis.ndc.nasa.gov/crgis/images/5/5b/L-66-6369.jpg

[32] http://www.armaghplanet.com/blog/whatever-happened-to-transient-lunar-phenomena.html

[33] https://spaceflight.nasa.gov/gallery/images/apollo/apollo15/html/s71-44666.html

[34] http://lroc.sese.asu.edu/posts/426

[35] http://bit.ly/2hSqnHx

[36] https://svs.gsfc.nasa.gov/3275

[37] http://www.thelivingmoon.com/43ancients/41Group_Lunar_FYEO/02files/FYEO_Lunar_02.html

[38] http://www.abovetopsecret.com/forum/thread467581/pg3

[39] https://science.nasa.gov/science-news/science-at-nasa/2014/13oct_moonvolcano/

[40] https://www.amazon.co.uk/Who-Built-Moon-Christopher-Knight-ebook/dp/B00IPHB3R4/

[41] https://www.abebooks.co.uk/servlet/SearchResults?cm_sp=plpafe-_-all-_-soft&an=don+wilson&bi=s&ds=5&n=&sortby=17&tn=mysterious+spaceship+moon

[42] http://www.auricmedia.net/wp-content/uploads/2015/01/Don_Wilson_Spaceship_Moon.pdf

[43] https://archive.org/details/SomebodyElseIsOnTheMoon

[44] https://crgis.ndc.nasa.gov/crgis/images/f/f7/L-67-7324.jpg

[45] http://bit.ly/2hTpCOv

[46] http://bit.ly/2hWmVvL

[47] https://nssdc.gsfc.nasa.gov/imgcat/hires/lo5_h168_2.gif

[48] https://the-moon.wikispaces.com/Vitello

[49] http://onebigmonkey.com/itburns/syfy/syfy1.html

[50] http://bit.ly/2A7WVHL

[51] http://the-moon.wikispaces.com/Boulder+Tracks

[52] http://bit.ly/2hT3i7B

[53] http://www.kmatthews.org.uk/face_on_mars/blair_cuspids.html

[54] http://www.astrosurf.com/lunascan/cuspids/Cuspid_LRO_Search.htm

[55] http://bit.ly/2A8Pm3D

[56] http://wms.lroc.asu.edu/lroc/view_lroc/LRO-L-LROC-3-CDR-V1.0/M159847595RC

[57] http://carlotto.us/newfrontiersinscience/Papers/v01n02c/v01n02c.pdf

[58] https://the-moon.wikispaces.com/Ukert

[59] https://tinyurl.com/urkertusgs

[60] http://bit.ly/2hTgZ6z

[61] http://www.lunar-captures.com/Domes_files/120103_domes_Ukert_Tar.jpg

[62] http://ser.sese.asu.edu/LO/lo3-hires.data.html

[63] http://ser.sese.asu.edu/LO/lo3-143-h3a.html

[64] https://timesofindia.indiatimes.com/india/Chandrayaan-beams-back-40000-images-in-75-days/articleshow/3979496.cms?

65 http://www.ufo-blogger.com/2009/01/indian-moon-mission-pictures-show.html

66 https://web.archive.org/web/20091012053414/http:/www.isro.org/Chandrayaan/images/1map_hysi.jpg

67 https://webapps.issdc.gov.in/CHBrowse/index.jsp

68 http://bit.ly/2hVwUBr

69 http://www.enterprisemission.com/mphotos.html

70 https://str.llnl.gov/etr/pdfs/06_94.1.pdf

71 https://www.nrl.navy.mil/clementine/clm/

72 http://www.thelivingmoon.com/43ancients/02files/Moon_Images_A28.html

73 https://www.youtube.com/watch?v=UOKnRjAWjh4

74 https://youtu.be/UOKnRjAWjh4?t=24m51s

75 http://wms.lroc.asu.edu/lroc/view_rdr_product/WAC_VIS_E235S1590_100MPP

76 http://lroc.sese.asu.edu/data/LRO-L-LROC-5-RDR-V1.0/LROLRC_2001/EXTRAS/BROWSE/WAC_COLOR_E235S1590/WAC_VIS_E235S1590_100MPP_RGB.TIF

77 https://dorkmission.blogspot.co.uk/2014/06/robert-morningstar-misinforms-public.html

78 http://www.carlotto.us/demos/googleEarthMarsCydoniaDemo.htm

79 http://nssdc.gsfc.nasa.gov/planetary/viking.html

80 https://ipfs.io/ipfs/QmXoypizjW3WknFiJnKLwHCnL72vedxjQkDDP1mXWo6uco/wiki/Cydonia_Mensae.html

81 https://mars.nasa.gov/resources/6279/

82 http://www.robertschoch.net/VD_CT_Findings_of_Life_on_Mars.htm

83 https://nssdc.gsfc.nasa.gov/image/planetary/mars/f070a13_processed.jpg

84 https://nssdc.gsfc.nasa.gov/photo_gallery/photogallery-mars.html

85 http://carlotto.us/martianenigmas/index.shtml

86 https://history.nasa.gov/SP-4212/ch9.html

87 https://books.google.co.uk/books?id=srxWAgAAQBAJ&pg=PA55&lpg=PA55&dq=sagan+astrology+mars+face++%22computer+enhancement+techniques+%22&source=bl&ots=4KhLMb5ygq&sig=T2cP0jHzON58HzJGUba43_TKffQ&hl=en&sa=X&ved=0ahUKEwip6qbL1_3XAhUrL8AKHQ_tD4QQ6AEIOTAG

88 https://nssdc.gsfc.nasa.gov/nmc/spacecraftDisplay.do?id=1992-063A

89 https://www.amazon.com/McDaniel-Report-Congressional-Responsibility-Investigating/dp/1556430884

90 https://www.bibliotecapleyades.net/marte/esp_marte_41.htm

91 http://carlotto.us/martianenigmas/Articles/aCommunicationsBreakthru.pdf

92 http://www.msss.com/mars_images/moc/4_6_98_face_release/face_w_craters.gif

93 http://citeseerx.ist.psu.edu/viewdoc/download?doi=10.1.1.594.8503&rep=rep1&type=pdf

94 http://spsr.utsi.edu/articles/jsefnl.htm

95 http://www.checktheevidence.com/cms/index.php?option=com_content&task=view&id=325&Itemid=60

96 http://tinyurl.com/911ftb

97 https://aas.org/obituaries/tom-c-van-flandern-1940-2009

98 https://www.bibliotecapleyades.net/sumer_anunnaki/esp_sumer_annunaki23.htm

99 http://solarviews.com/eng/miranda.htm

100 https://photojournal.jpl.nasa.gov/catalog/PIA18185

101 https://www.bibliotecapleyades.net/marte/esp_marte_29.htm

102 http://www.dailymail.co.uk/sciencetech/article-2998835/Were-Martians-wiped-nuclear-bomb-Physicist-present-new-evidence-bizarre-theory-Nasa-conference.html

103 http://spsr.utsi.edu/articles/EvidenceforaLargeAnomalousNuclearExplosionsinMarsPast.pdf

104 https://www.youtube.com/watch?v=zd6xa5VivaE

105 http://www.thunderbolts.info/tpod/2011/arch11/110420wargod.htm

106 http://www.skeptiko-forum.com/threads/dr-john-brandenburg-responds-to-musselwhite-and-electric-universe.3401/

107 http://www.msss.com/mars_images/moc/extended_may2001/face/index.html

108 http://www.21stcenturyradio.com/articles/0605013.html

109 https://science.nasa.gov/science-news/science-at-nasa/2001/ast24may_1

110 http://www.msss.com/mars_images/moc/extended_may2001/face/face_E03-00824_proc_50perc.gif

111 https://www.youtube.com/watch?v=JPJRpBWuGbY

112 https://www.matrixwissen.de/index.php?option=com_content&view=article&id=850:disclosure-project-press-conference-2001-en&catid=211:a-new-reality-3-en&lang=en&Itemid=520

113 http://www.encyclopedia.com/arts/educational-magazines/luckman-michael-c

114 https://www.youtube.com/watch?v=L777RhL_Fz4

115 http://www.maxtheknife.com/cruxofcydoniaWHITE.pdf

116 https://www.amazon.co.uk/Cydonia-Codex-Reflections-Mars/dp/1583941215

117 http://viewer.mars.asu.edu/planetview/inst/themis/V03814003

118 http://www.enterprisemission.com/paper_1/mars_paper.php

119 http://www.esa.int/Our_Activities/Space_Science/Mars_Express/Cydonia_-_the_face_on_Mars

120 http://www.esa.int/Our_Activities/Space_Science/Mars_Express/Cydonia_s_Face_on_Mars_in_3D_animation

121
http://www.esa.int/var/esa/storage/images/esa_multimedia/images/2006/09/face_on_mars_in_cydonia_region_perspective2/9772500-3-eng-GB/Face_on_Mars_in_Cydonia_region_perspective.tif

122 http://www.esa.int/var/esa/storage/images/esa_multimedia/images/2006/09/face_on_mars_in_cydonia_region/9624502-3-eng-GB/Face_on_Mars_in_Cydonia_region_large.jpg

123 http://thecydoniainstitute.com/The-ESA-Team-Creates-a-Horned-Face.php

124 http://www.aulis.com/mars.htm

125 http://sci.esa.int/science-e-media/img/4c/Cydonia_Region.jpg

126 https://www.uahirise.org/PSP_003234_2210

127 http://hiroc.lpl.arizona.edu/images/PSP/PSP_003234_2210/PSP_003234_2210_RED.browse.jpg

128 http://tardis.wikia.com/wiki/Pyramids_of_Mars_(TV_story)

129 http://www.msss.com/mars_images/moc/4_5_00_cydonia/

130 http://users.starpower.net/etorun/pyramid/

131 http://www.msss.com/mars_images/moc/2003/09/15/

132 http://www.msss.com/mars_images/moc/2003/09/15/DandM100.gif

133 http://www.gigapan.com/galleries/13462/gigapans/190739

[134] http://www.msss.com/mars_images/moc/4_5_00_cydonia/full_res_images/sp1-25803d.gif

[135] http://www.msss.com/mars_images/moc/4_5_00_cydonia/full_res_images/m09-05394d.gif

[136] http://www.msss.com/mars_images/moc/4_5_00_cydonia/full_res_images/m04-01903d.gif

[137] https://www.gimp.org/downloads/

[138] https://picasa.en.uptodown.com/windows

[139] http://mars-news.de/mr9/4205-78.html

[140] https://archive.org/stream/Cosmos-CarlSagan/cosmos-sagan_djvu.txt

[141] http://viewer.mars.asu.edu/planetview/inst/ctx/P03_002318_1961_XN_16N198W

[142] http://www.mbgnet.net/fresh/lakes/oxbow.htm

[143] https://www.youtube.com/watch?time_continue=5&v=c2y9E-PiKjE

[144] http://themis-data.asu.edu/planetview/inst/themis/I03502047

[145] http://davidpratt.info/mars-life.htm

[146] http://carlotto.us/demos/geCydoniaImages/35A74.jpg

[147] http://www.msss.com/mars_images/moc/4_5_00_cydonia/full_res_images/m03-00766d.gif

[148] http://carlotto.us/martianenigmas/Articles/April_2000/Tholus360.gif

[149] http://asimov.msss.com/moc_gallery/m13_m18/images/M18/M1800606.html

[150] http://www.enterprisemission.com/europe01.html

[151] http://www.avebury-web.co.uk/avebury_now.html

[152] http://www.rexresearch.com/smith/newsci.htm

[153] http://www.treurniet.ca/Smith/SmithCoil.htm

[154] http://www.zipcon.net/~swhite/docs/astronomy/Angular_Momentum.html

[155] http://www.astronoo.com/en/articles/angular-momentum.html

[156] https://www.brucedepalma.com/

[157] http://www.rexresearch.com/smith/smith2.htm

[158] https://voyager.jpl.nasa.gov/galleries/images-voyager-took/neptune/

[159] http://www.halexandria.org/dward760.htm

[160] https://www.youtube.com/watch?v=UnUREICzGc0

[161] http://www.esa.int/spaceinimages/Images/2006/01/Close-up_perspective_view_of_the_sulphate_mountain_-_looking_east

[162] http://sci.esa.int/mars-express/38703-juventae-chasma/

[163] http://esamultimedia.esa.int/images/marsexpress/054co01JuventaeChasma_H.jpg

[164] http://esamultimedia.esa.int/images/marsexpress/0563D101JuventaeChasma_H.jpg

[165] http://mcdonaldmorrissey.com/barrick-goldstrike-mines-inc-elko-nevada/

[166] https://www.redzaustralia.com/2013/04/unnatural-attractions-the-super-pit-kalgoorlie-boulder-western-australia/

[167] https://hirise.lpl.arizona.edu/ESP_020794_1860

[168] http://www.ufo-blogger.com/2014/01/japan-kofun-era-tomb-structure-found-on-mars.html

[169] http://dailyglimpsesofjapan.blogspot.co.uk/2012/07/kofun-ancient-japanese-tombs.html

[170] https://ida.wr.usgs.gov/html/sp2430/sp243004.html

171 https://www.amazon.co.uk/Secret-Mars-Alien-Connection-Craig/dp/0992605342

172 https://ida.wr.usgs.gov/html/e10004/e1000462.html

173 https://ida.wr.usgs.gov/fullres/divided/e10004/e1000462a.jpg

174 http://www.msss.com/mars_images/moc/8_2002_releases/incacity/das8044333sub.gif

175 https://ida.wr.usgs.gov/html/m04002/m0400291.html

176 https://ida.wr.usgs.gov/fullres/divided/m04002/m0400291a.jpg

177 http://www.badastronomy.com/info/whois.html

178 http://edition.cnn.com/2004/TECH/space/03/17/alien.debunk/index.html

179 http://www.xfacts.com/

180 https://www.secretmars.com/

181 http://ida.wr.usgs.gov/html/m15012/m1501228.html

182 http://asimov.msss.com/moc_gallery/m13_m18/images/M15/M1501228.html

183 https://ida.wr.usgs.gov/html/m15012/m1501228.html

184 http://palermoproject.com/lowell2004/grandcentral.htm

185 https://disneyworld.disney.go.com/en_GB/attractions/epcot/spaceship-earth/

186 http://web.archive.org/web/20130522122953/http:/eclipseedge.org/msgboard/topic.asp?TOPIC_ID=1025

187 http://web.archive.org/web/20090511150126/http:/hirise.lpl.arizona.edu/PSP_007230_2170

188 http://hirise.lpl.arizona.edu/PSP_007230_2170

189 https://hirise-pds.lpl.arizona.edu/PDS/EXTRAS/RDR/PSP/ORB_007200_007299/PSP_007230_2170/PSP_007230_2170_RED.abrowse.jpg

190 https://www.youtube.com/watch?v=tRV1e5_tB6Y&feature=youtu.be&t=1h20m52s

191 https://ida.wr.usgs.gov/fullres/divided/m15012/m1501228b.jpg

192 http://www.richplanet.net/rovers.php

193 https://nssdc.gsfc.nasa.gov/nmc/SpacecraftQuery.jsp

194 https://nssdc.gsfc.nasa.gov/nmc/spacecraftDisplay.do?id=1996-068A

195 https://nssdc.gsfc.nasa.gov/nmc/spacecraftDisplay.do?id=2003-032A

196 http://www.enterprisemission.com/spirit2.htm

197 https://mars.jpl.nasa.gov/MPF/parker/TwnPks_RkGdn_left_high.jpg

198 http://www.enterprisemission.com/Path-sphinx.html

199 https://hirise.lpl.arizona.edu/PSP_001890_1995

200 https://mars.jpl.nasa.gov/msl/multimedia/raw/?rawid=0173MR0926020000E1_DXXX&s=173

201 https://www.universetoday.com/99890/scientist-explains-the-weird-shiny-thing-on-mars/

202 https://nssdc.gsfc.nasa.gov/planetary/factsheet/earthfact.html

203 https://nssdc.gsfc.nasa.gov/planetary/factsheet/marsfact.html

204 http://www.marsanomalies.com/metronome

205 https://mars.nasa.gov/mer/gallery/all/2/p/1402/2P250825588EFFAW9DP2432R1M1.HTML

206 http://www.alienstudy.com/2012/12/metronome/

[207] http://pancam.sese.asu.edu/images/Sol1401A_P2431_1_L257atc.jpg

[208] https://mars.jpl.nasa.gov/mer/gallery/all/2/n/033/2N129300816EFF0327P1730L0M1.HTML

[209] https://mars.jpl.nasa.gov/msl-raw-images/proj/msl/redops/ods/surface/sol/00665/opgs/edr/ncam/NRB_456540999EDR_F0361146NCAM00268M_.JPG

[210] https://mars.nasa.gov/msl/multimedia/raw/?rawid=0003ML0000037000E1_DXXX&s=3

[211] https://mars.jpl.nasa.gov/msl-raw-images/msss/00882/mhli/0882MH0003900000302461E01_DXXX.jpg

[212] https://mars.jpl.nasa.gov/mer/gallery/all/2/n/581/2N177950967EFFADNDP0645R0M1.HTML

[213] http://www.sitchin.com/

[214] https://mars.nasa.gov/mer/gallery/all/2/p/513/2P171912249EFFAAL4P2425R1M1.HTML

[215] https://www.britannica.com/place/Mars-planet

[216] https://nssdc.gsfc.nasa.gov/imgcat/html/object_page/vl2_p21873.html

[217] http://www.esa.int/Our_Activities/Space_Science/Mars_Express/Water_ice_in_crater_at_Martian_north_pole

[218] https://phys.org/news/2012-09-temperatures-gale-crater-higher.html

[219] https://www.space.com/16907-what-is-the-temperature-of-mars.html

[220] http://news.bbc.co.uk/1/hi/sci/tech/3387903.stm

[221] https://pds-imaging.jpl.nasa.gov/data/mer/mer1po_0xxx/browse/sol0257/edr/1p150975152eff36cbp2693l2m1.img.jpg

[222] https://mars.nasa.gov/mer/gallery/all/1/f/119/1F138744391EFF2809P1214R0M1.JPG

[223] http://mars.nasa.gov/mer/gallery/all/1/p/123/1P139113540EFF2811P2535L7M1.JPG

[224] https://mars.nasa.gov/mer/mission/spacecraft_instru_mossbr.html

[225] https://web.archive.org/web/20080926001046/http:/xenotechresearch.com:80/newflow1.htm

[226] http://apod.nasa.gov/apod/ap000626.html

[227] https://www.nasa.gov/mission_pages/mars/news/mgs-20061206.html

[228] http://www.nasa.gov/mission_pages/MRO/news/mro20110804.html

[229] http://www.bbc.co.uk/news/av/science-environment-34385359/mars-satellite-hints-at-liquid-water

[230] http://www.breitbart.com/big-government/2015/09/29/ridley-scott-confirms-nasa-timed-mars-water-find-to-boost-damons-martian-movie/

[231] http://www.gillevin.com/mars.htm

[232] http://www.spie.org/newsroom/levin-video?SSO=1

[233] http://www.gillevin.com/Mars/Reprint118-Spie2001-oxides_files/Reprint118-Spie2001-oxides.htm

[234] http://www.inchem.org/documents/ehc/ehc/ehc89.htm

[235] http://news.bbc.co.uk/1/hi/sci/tech/4295475.stm

[236] http://www.belfasttelegraph.co.uk/breakingnews/breakingnews_world/breakingnews_world_northamerica/nasa-methane-clouds-could-be-evidence-of-life-on-mars-28462352.html

[237] https://science.nasa.gov/science-news/science-at-nasa/2009/15jan_marsmethane

[238] http://web.archive.org/web/20130106150152/http:/www.wired.co.uk/news/archive/2012-12/04/mars-curiosity-soil-sample-results

[239] http://www.dailymail.co.uk/sciencetech/article-2426424/NASA-says-NO-life-Mars-Curiosity-rover-discovered-clues-atmosphere-supports-living-things.html

[240] http://www.wired.co.uk/article/nasa-curiosity-methane-mars

[241] https://science.nasa.gov/science-news/science-at-nasa/2014/16dec_methanespike/

[242] http://www.ufo-blogger.com/2013/02/mars-fossilized-spine-nasa-curiosity.html

[243] https://www.youtube.com/watch?v=ezXMejUUA5Y

[244] https://marsmobile.jpl.nasa.gov/msl/multimedia/raw/?rawid=0109MR0684021000E1_DXXX&s=109

[245] https://www.scribd.com/doc/289291021/The-Living-Rocks-of-Mars

[246] https://www.britannica.com/science/stromatolite

[247] http://www.fossilmall.com/Stonerelic/stromatolite/Stro17/Stro17.htm

[248] https://mars.nasa.gov/mer/gallery/all/1/m/3064/1M400198734EFFBVM5P2956M2M2.JPG

[249] https://mars.nasa.gov/mer/gallery/all/1/m/034/1M131201538EFF0500P2933M2M1.HTML

[250] https://www.coasttocoastam.com/guest/shults-iii-charles/6327

[251] https://www.youtube.com/watch?v=37-QDhQs9AY

[252] https://www.coasttocoastam.com/show/2005/08/21

[253] http://www.shultslaboratories.com/AFHG2MPR.htm

[254] http://www.shultslaboratories.com/SL2D007.htm

[255] https://mars.nasa.gov/mer/gallery/all/1/m/505/1M173017292EFF55VWP2956M2M1.HTML

[256] http://www.shultslaboratories.com/SL2D010.htm

[257] https://mars.nasa.gov/mer/gallery/all/2/p/343/2P156815447EFFA201P2574L6M1.HTML

[258] http://www.shultslaboratories.com/SL2D004.htm

[259] https://mars.nasa.gov/mer/gallery/all/1/m/507/1M173191393EFF55W4P2956M2M1.HTML

[260] http://www.shultslaboratories.com/SL2D006.htm

[261] https://www.britannica.com/animal/copepod

[262] https://mars.nasa.gov/mer/gallery/all/1/m/239/1M149401056EFF35CYP2977M2M1.HTML

[263] http://www.shultslaboratories.com/SL2K001.htm

[264] https://mars.nasa.gov/mer/gallery/all/2/p/007/2P126991583EFF0205P2530L5M1.JPG

[265] http://cosmology.com/MarsCurrentPastLife.html

[266] https://www.britannica.com/biography/Arthur-C-Clarke

[267] http://asimov.msss.com/moc_gallery/m07_m12/nonmaps/M08/M0804688.gif

[268] http://asimov.msss.com/moc_gallery/m07_m12/images/M08/M0804688.html

[269] http://web.archive.org/web/20131119080219/http:/www.arthurcclarke.net/?interview=7

[270] https://www.popsci.com/military-aviation-space/article/2001-12/banyan-trees-mars

[271] http://www.msss.com/mars_images/moc/8_10_99_releases/moc2_166/index.html

[272] http://www.msss.com/mars_images/moc/8_10_99_releases/moc2_166/moc2_166b_msss.gif

[273] https://mars.jpl.nasa.gov/mgs/msss/camera/images/dune_defrost_6_2001/

[274] https://mars.jpl.nasa.gov/mgs/msss/camera/images/dune_defrost_6_2001/e05-762.gif

[275] http://www.treesonmars.org/rings.html

[276] https://www.lpi.usra.edu/

[277] https://www.lpi.usra.edu/meetings/lpsc2004/pdf/1027.pdf

[278] https://www.youtube.com/watch?v=8AtC4Woxi7M

[279] http://asimov.msss.com/moc_gallery/ab1_m04/images/SP253807.html

[280] http://asimov.msss.com/moc_gallery/ab1_m04/nonmaps/SP253807.gif

[281] https://setiathome.berkeley.edu/forum_thread.php?id=20513&postid=172221

[282] http://www.treesonmars.org/home-1.html

[283] https://www.space.com/7775-strange-mars-photo-includes-tantalizing-tree-illusion.html

[284] http://www.telegraph.co.uk/news/science/space/6979855/Nasa-photographs-trees-on-Mars.html

[285] https://www.express.co.uk/news/weird/671136/TREES-FOUND-ON-MARS-Extraordinary-image-shows-towering-plants-on-the-Red-Planet

[286] https://www.wsj.com/articles/SB10001424052748703791904576076360797960364

[287] https://apod.nasa.gov/apod/ap100119.html

[288] https://hirise.lpl.arizona.edu/PSP_007962_2635

[289] https://www.youtube.com/watch?v=zhfSiJeQf58

[290] http://phoenix.lpl.arizona.edu/imageCategories_lander.php

[291] https://www.britannica.com/animal/tardigrade

[292] https://www.livescience.com/57985-tardigrade-facts.html

[293] https://www.youtube.com/watch?v=LUHZ-aMbZHw

[294] https://www.scientificamerican.com/article/martian-soil-fit-for-earthly-life/

[295] http://www.enterprisemission.com/_articles/04-13-2004_Methane_on_Mars/Methane_on_Mars_1.htm

[296] http://www.earthfiles.com/news/news.cfm?ID=650&category=Science

[297] http://web.archive.org/web/20041020113215/http:/www.dlr.de:80/mars-express/images/230104

[298] http://web.archive.org/web/20041111122642/http:/www.dlr.de:80/mars-express/images/230104/gusev_3_ColorComplete_900.jpg

[299] http://sci.esa.int/science-e-media/img/cc/reull_vallis_river_channel.jpeg

[300] http://www.enterprisemission.com/colors2.htm

[301] http://news.bbc.co.uk/1/hi/sci/tech/1913228.stm

[302] https://www.digitaltrends.com/mobile/camera-phone-history/

[303] https://www.atoptics.co.uk/atoptics/sunsets.htm

[304] https://www.youtube.com/watch?v=SkqBT6d0VWU

[305] http://www.enterprisemission.com/colors.htm

[306] https://www.amazon.co.uk/dp/B004Q7CMFY/

[307] https://photojournal.jpl.nasa.gov/catalog/PIA00563

[308] https://youtu.be/zg1NUs5Txsl?t=1m57s

[309] http://www.americaspace.com/2016/07/19/viking-remembered-celebrating-the-40th-anniversary-of-the-first-search-for-life-on-mars/

[310] https://pds-imaging.jpl.nasa.gov/data/vl1_vl2-m-lcs-2-edr-v1.0/vl_0001/browse/html/b1xx/12b166gn.htm

[311] http://mars-news.de/color/blue.html

[312] http://mars-news.de/color/12B069.jpg

313 http://www.gillevin.com/Mars/Reprint125_files/Reprint125-SPIE-2003-Color-Paper.htm

314 http://www.gillevin.com/Mars/5555-29.PDF

315 http://www.gillevin.com/Mars/Reprint87-color-files/colorReprint87.htm

316 http://www.goroadachi.com/etemenanki/mars-hiddencolors.htm

317 http://www.abovetopsecret.com/forum/thread30048/pg1

318 http://www.gillevin.com/Mars/5555-30.PDF

319 https://mars.jpl.nasa.gov/mer/gallery/all/1/p/097/1P136800137ESF2002P2559L2M1.JPG

320 http://sci.esa.int/mars-express/50840-mars-facing-side-of-phobos/

321 https://www.britannica.com/place/Phobos-moon-of-Mars

322 https://www.youtube.com/watch?v=XcHkXgesG0M

323 https://archive.org/details/SaganIL

324 http://www.presidentialufo.com/old_site/eisenhow5.htm

325 http://www.fampeople.com/cat-fred-singer_4

326 https://www.youtube.com/watch?v=_grnu3u149o

327 http://www.cufos.org/UFOI_and_Selected_Documents/UFOI/019 JAN-FEB 1963.pdf

328 https://www.gutenberg.org/files/829/829-h/829-h.htm

329 https://nssdc.gsfc.nasa.gov/imgcat/html/mission_page/MS_Viking_1_Orbiter_page1.html

330 https://apod.nasa.gov/apod/ap100317.html

331 http://sci.esa.int/mars-express/31031-phobos/

332 http://onlinelibrary.wiley.com/doi/10.1029/2009GL041829/abstract

333 https://photojournal.jpl.nasa.gov/catalog/PIA14894

334 https://photojournal.jpl.nasa.gov/catalog/?IDNumber=PIA15678

335 https://youtu.be/E8aqWgeO8oo?t=1h23m00s

336 https://photojournal.jpl.nasa.gov/jpegMod/PIA15824_modest.jpg

337 https://nssdc.gsfc.nasa.gov/nmc/spacecraftDisplay.do?id=1988-058A

338 http://www.angelfire.com/zine/UFORCE/page72.html

339 http://forgetomori.com/2009/ufos/phobos-2-a-bloody-soviet-close-encounter/

340 http://www.planetary.org/explore/resource-library/data/phobos-2-vsk-data.html

341 https://www.bibliotecapleyades.net/sitchin/genesisrevisto/genrevisit12.htm

342 http://www.esa.int/Our_Activities/Space_Science/Mars_Express/Light_and_shadow_on_the_surface_of_Mars

343 https://www.youtube.com/watch?v=65zZpRgrRTg

344 http://palermoproject.com/Mars_Anomalies/PhobosAnomalies1.html

345 https://ida.wr.usgs.gov/html/orb_0551/55103.html

346 https://d.docs.live.net/0c6ebe1d913dd93a/ida.wr.usgs.gov/fullres/divided/orb_0551/55103h.jpg

347 https://youtu.be/2QFgdo3BtyQ

348 https://www.youtube.com/watch?v=r_IKEJaieGc

349 http://www.dailymail.co.uk/sciencetech/article-1204254/Has-mystery-Mars-Monolith-solved.html

350 http://i.dailymail.co.uk/i/pix/2009/08/04/article-1204254-05F2E104000005DC-903_634x264.jpg

351 http://apod.lunexit.it/search.php?q=monolith

352 http://www.enterprisemission.com/Phobos2.html

353 https://photojournal.jpl.nasa.gov/catalog/PIA10369

354 https://static.uahirise.org/images/2008/details/phobos/PSP_007769_9010_IRB.jpg

355 http://www.enterprisemission.com/Phobos-monolith-loc.jpg

356 https://youtu.be/Cn3URo5KkdM

357 https://www.marspages.eu/index.php?page=512

358 https://hirise.lpl.arizona.edu/PSP_009342_1725

359 http://www.telegraph.co.uk/news/science/space/5981624/Mars-monolith-fuels-theories-of-alien-life.html

360 http://www.bbc.co.uk/earth/story/20160923-there-is-a-huge-monolith-on-phobos-one-of-marss-moons

361 http://www.dailymail.co.uk/news/article-501316/The-Pope-condemns-climate-change-prophets-doom.html

362 http://www.checktheevidence.com/audio/Rush-Limbaugh-March 4-2004-The Mars-(Gore)-Report.mp3

363 https://trekmovie.com/2008/04/09/shatner-claims-to-know-about-life-on-mars/

364 https://youtu.be/It9f7dxCh6Y

365 https://youtu.be/JW6cJuKDUsU?t=49m15s

366 https://www.bibliotecapleyades.net/disclosure/briefing/disclosure12.htm

367 https://www.youtube.com/watch?v=8pa3ZT6QHjo

368 http://exopolitics.org/jump-room-to-mars-did-cia-groom-obama-basiago-as-future-presidents/

369 http://www.stillnessinthestorm.com/2017/09/transcript-of-corey-goode-presentation-life-on-mars-sedona-cosmic-awakening-conference-april-21st-2017.html

370 https://www.gaia.com/article/randy-cramer-mars-defense-force

371 http://www.themarsrecords.com/

372 http://www.indiana.edu/~p1013447/dictionary/circrhyt.htm

373 https://news.harvard.edu/gazette/story/1999/07/human-biological-clock-set-back-an-hour/

374 http://www.richplanet.net/starship_main.php?ref=192&part=1

375 http://www.enterprisemission.com/

376 http://www.aulis.com/

377 https://archive.org/details/NASAMoonedAmericaByRalphRene1994237P_201604

378 https://www.scribd.com/doc/156629102/We-Never-Went-to-the-Moon-By-Bill-Kaysing

379 http://www.aulis.com/films.htm

380 http://www.moonfaker.com/videos.php

381 http://www.checktheevidence.com/pdf/Apollo-RemovingTruthsProtectiveLayers-Booklet.pdf

382 https://www.youtube.com/watch?v=5KsH2M4m4zM

383 https://www.youtube.com/watch?v=Znyx2gTh3HU

384 https://spaceflight.nasa.gov/gallery/images/apollo/apollo16/lores/as16-107-17442.jpg

385 https://www.hq.nasa.gov/office/pao/History/alsj/a16/AS16-107-17435HR.jpg

386 http://airandspace.si.edu/explore-and-learn/multimedia/detail.cfm?id=5496

387 http://www.spacegrant.hawaii.edu/class_acts/MoonFacts.html

388 http://www.aulis.com/stereoparallax.htm

389 http://www.nasa.gov/mission_pages/LRO/multimedia/lroimages/lroc_200911109_apollo11.html

390 http://www.americanmoon.org/NationalGeographic/index.htm

391 https://www.space.com/20865-soviet-moon-rover-lunokhod-laser.html

392 https://www.youtube.com/watch?v=0eDaQo29E-w

393 http://news.bbc.co.uk/1/hi/8226075.stm

394 http://mikebara.blogspot.co.uk/2008/03/why-moon-hoax-conspiracy-is-crock-of.html

395 http://www.imdb.com/title/tt0077294/

396 http://www.checktheevidence.com/pdf/Richplanet_Rovers_1.pdf

397 https://photojournal.jpl.nasa.gov/catalog/PIA16204

398 https://mars.nasa.gov/mer/gallery/all/2/m/386/2M160631572EFFA2K1P2936M2M1.HTML

399 https://www.youtube.com/watch?v=89YfPWTpWN0

400 https://www.youtube.com/watch?v=FVzfDZlEwaU

401 https://youtu.be/QH7vN3oO1Zw?t=3m17s

402 https://www.space.com/16946-mars-rover-landing-seen-from-space.html

403 https://www.space.com/24227-mars-rover-curiosity-photos-from-space.html

404 https://www.space.com/28475-curiosity-mars-rover-space-photo.html

405 http://www.checktheevidence.com/pdf/Science Fiction and the Hidden Global Agenda - Carl James - 1st Ed - 2014.pdf

406 https://saturn.jpl.nasa.gov/mission/grand-finale/cassini-end-of-mission-timeline/

407 https://www.britannica.com/biography/Christiaan-Huygens

408 http://sci.esa.int/cassini-huygens/36378-disr-image-of-titan-s-surface/

409 http://aasnova.org/2015/08/21/an-explanation-for-saturns-hexagon/

410 http://www.planetary.org/multimedia/space-images/saturn/saturn-north-polar-hexagon-vims-animation.html

411 https://www.sciencealert.com/saturn-s-mysterious-hexagon-has-changed-from-blue-to-gold-and-no-one-knows-why

412 https://saturn.jpl.nasa.gov/science/magnetosphere/

413 http://www.rolfolsenastrophotography.com/Astrophotography/Solar-System/i-vjMHSxz/A

414 http://podcast.sjrdesign.net/files/070_RingmakersOfSaturn.pdf

415 https://www.youtube.com/watch?v=RqkohQwY-c0

416 https://www.youtube.com/watch?v=kPgdyWE7lbo

417 https://www.universetoday.com/81774/bright-white-storm-raging-on-saturn/

418 https://www.youtube.com/watch?v=ibT4SFNcGcY

419 https://saturn.jpl.nasa.gov/images/casJPGFullS31/N00084960.jpg

420 https://saturn.jpl.nasa.gov/images/casJPGFullS31/N00084964.jpg

421 https://saturn-archive.jpl.nasa.gov/photos/raw/rawimagedetails/?imageID=114142

422 https://saturn.jpl.nasa.gov/images/casJPGFullS35/W00039344.jpg

423 https://saturn.jpl.nasa.gov/images/casJPGFullS35/W00039360.jpg

424 https://saturn.jpl.nasa.gov/raw_images/164671/

425 https://photojournal.jpl.nasa.gov/catalog/PIA11668

426 http://www.planetary.org/blogs/emily-lakdawalla/2014/07010001-ringmoons-shepherds.html

427 https://www.britannica.com/place/Saturn-planet/Orbital-and-rotational-dynamics

428 http://solarviews.com/eng/atlas.htm

429 https://photojournal.jpl.nasa.gov/catalog/PIA08405

430 https://www.nasa.gov/feature/jpl/cassini-sees-flying-saucer-moon-atlas-up-close

431 http://www.planetary.org/blogs/guest-blogs/2017/20170313-three-discoveries-of-pan.html

432 http://www.enchantedlearning.com/subjects/astronomy/planets/saturn/saturnmoons.shtml

433 https://saturn.jpl.nasa.gov/raw_images/103924/

434 https://www.jpl.nasa.gov/news/news.php?feature=6770

435 http://www.skyandtelescope.com/astronomy-news/welcome-pan-saturns-ravioli-shaped-moon/

436 http://enterprisemission.com/moon1.htm

437 https://photojournal.jpl.nasa.gov/catalog/PIA00348

438 http://www.planetary.org/blogs/emily-lakdawalla/2012/3389.html

439 https://www.jpl.nasa.gov/spaceimages/details.php?id=PIA06100

440 https://photojournal.jpl.nasa.gov/catalog/PIA06166

441 https://photojournal.jpl.nasa.gov/jpegMod/PIA06166_modest.jpg

442 http://www.enterprisemission.com/moon2.htm

443 https://photojournal.jpl.nasa.gov/catalog/PIA06146

444 https://youtu.be/PKNSpQPyxYE?t=23m55s

445 https://photojournal.jpl.nasa.gov/catalog/PIA08384

446 https://photojournal.jpl.nasa.gov/jpeg/PIA08384.jpg

447 https://saturn.jpl.nasa.gov/galleries/raw-images?order=earth_date+desc&per_page=50&page=8&begin_date=2006-12-31&end_date=2007-12-31&targets%5B%5D=IAPETUS

448 https://saturn.jpl.nasa.gov/images/casJPGFullS33/W00035179.jpg

449 https://saturn.jpl.nasa.gov/images/casJPGFullS33/W00035187.jpg

450 https://saturn.jpl.nasa.gov/images/casJPGFullS33/W00035194.jpg

451 https://theconversation.com/what-cassinis-mission-revealed-about-saturns-known-and-newly-discovered-moons-83430

452 http://onlinelibrary.wiley.com/doi/10.1029/2011JE004010/epdf

453 https://saturn.jpl.nasa.gov/resources/4772/

454 http://theconversation.com/what-cassinis-mission-revealed-about-saturns-known-and-newly-discovered-moons-83430

455 http://www.sciencedirect.com/science/article/pii/S0019103509003303

456 http://www.abovetopsecret.com/forum/thread499521/pg1

457 https://www.youtube.com/watch?v=NW9r17nEpSc

458 http://www.checktheevidence.co.uk/cms/index.php?option=com_content&task=view&id=50&Itemid=59

459 http://road.cc/content/feature/171115-pros-and-cons-carbon-fibre-wheels

460 https://www.britannica.com/science/fullerene

461 http://www.enterprisemission.com/moon6.htm

462 https://www.youtube.com/watch?v=QtlXAOoJp10

463 https://www.techsupportalert.com/content/celestia.htm-0

464
http://web.archive.org/web/20050801000000*/http:/sohowww.nascom.nasa.gov/data/realtime/javagif/gifs/20050102_1230_c2.
gif

465 https://lasco-www.nrl.navy.mil/index.php?p=content/handbook/hndbk_5

466 https://lasco-www.nrl.navy.mil/index.php?p=content/handbook/hndbk_6

467 https://sohowww.nascom.nasa.gov/about/docs/SOHO_Fact_Sheet.pdf

468 https://sohowww.nascom.nasa.gov/about/orbit.html

469 https://map.gsfc.nasa.gov/mission/observatory_l2.html

470 https://sohowww.nascom.nasa.gov/hotshots/2003_01_17/

471 https://sohowww.nascom.nasa.gov/hotshots/2003_01_17/NewspaperSOHOsmall.jpg

472 http://news.bbc.co.uk/1/hi/england/2662787.stm

473 http://www.nbcnews.com/id/3077760/ns/technology_and_science-space/t/scientists-show-how-make-ufo/

474 http://news.bbc.co.uk/1/hi/england/2662059.stm

475 http://solar.bnsc.rl.ac.uk/press/007/007.html

476 https://spacecentre.co.uk/

477 http://jsoc1.bnsc.rl.ac.uk/

478 http://www.stfc.ac.uk/about-us/where-we-work/rutherford-appleton-laboratory/

479 http://www.harwellcampus.com/about/about-harwell/

480 http://www.checktheevidence.com/cms/index.php?option=com_content&task=view&id=132&Itemid=59

481 http://www.checktheevidence.com/images/index.php?dir=SOHOAnomalies/

482 https://www.nrl.navy.mil/content_images/horizon.pdf

483 https://www.youtube.com/watch?v=uKlANT7gemg

484 http://www.imdb.com/title/tt0064519/

485 https://visibleearth.nasa.gov/view.php?id=49316

486 https://www.nasa.gov/mission_pages/sunearth/science/solar-rotation.html

487 https://www.amazon.co.uk/Complete-Earth-Chronicles/dp/1591432014/

488 http://www.sitchinstudies.com/the-sumerian-solar-system.html

489 http://www.sitchinstudies.com/nibiru.html

490 http://www.iau.org/static/archives/images/screen/iau0603a.jpg

491 https://www.youtube.com/watch?v=troqVst56eg

492 http://www.northern-stars.com/solar_system_info.htm

493 http://www.express.co.uk/news/science-technology/467077/Huge-planet-ten-times-bigger-than-Earth-could-be-orbiting-
Sun-at-edge-of-solar-system

494 http://earthsky.org/space/planet-nibiru-is-not-real

495 https://www.britannica.com/topic/occultation

496 http://xfacts.com/sumerian_culture.html

497 http://www.zetatalk.com/

498 http://planet-x.150m.com/where.html

499 http://www.zetatalk.com/index/pufoin2.htm

500 http://www.zetatalk.com/newsletr/issue167.htm

501 http://www.jmccanneyscience.com/Bio.HTM

502 http://web.archive.org/web/20080710062337/http:/planetxforecast.com/

503 http://yowusa.com/2017/12/planet-x-nibiru-has-arrived-no-4/

504 https://www.express.co.uk/news/weird/876285/Nibiru-may-2018-planet-x-news-world-war-3-wormwood-orbit-space-earthquakes-volcanoes

505 http://www.cosmophobia.org/nibiru/index.html

506 http://www.open.ac.uk/people/ghm2

507 https://ntrs.nasa.gov/archive/nasa/casi.ntrs.nasa.gov/19640053196.pdf

508 https://ia902707.us.archive.org/12/items/nasa_techdoc_19640053194/19640053194.pdf

509 https://www.youtube.com/watch?v=656kMTDA0Gk

510 http://www.lloydpye.com/

511 http://www.starchildproject.com/

512 http://www.drjudywood.com/

513 https://www.washingtonpost.com/news/morning-mix/wp/2015/02/24/how-stephen-hawking-survived-longer-than-possibly-any-other-als-patient/

514 https://www.theguardian.com/science/2010/apr/30/stephen-hawking-right-aliens

515 http://www.checktheevidence.com/Disclosure/PDF_Documents/COMETA_part1.pdf

516 http://milesmathis.com/hawk3.pdf

517 https://www.wired.co.uk/article/brian-cox-interview-alien-life-climate-change-donald-trump

518 https://www.youtube.com/watch?v=1aXuQ9Dg2gE

519 http://www.iflscience.com/space/brian-cox-explains-why-we-havent-seen-aliens-yet-and-it-isnt-pretty/

520 https://www.csicop.org/si/show/universe_and_carl_sagan

521 https://www.goodreads.com/book/show/61666.Contact

522 http://www.imdb.com/title/tt0118884/

523 https://archive.org/stream/DarkMissionTheSecretHistoryOfNASA_201611/Richard Hoagland/Dark Mission - The Secret History Of NASA_djvu.txt

524 http://www.project1947.com/shg/mccarthy/chap06.html

525 http://files.ncas.org/condon/text/contents.htm

526 https://archive.org/details/scienceindefault

527 http://www.stantonfriedman.com/index.php?ptp=articles&fdt=2004.04.15&prt=2

528 https://www.youtube.com/watch?v=avb6i9QMo5g

[529] http://ufoupdateslist.com/2009/may/m06-004.shtml

[530] https://nsarchive2.gwu.edu/NSAEBB/NSAEBB479/docs/EBB-Moon02.pdf

[531] https://www.youtube.com/watch?v=WhW3_L8Vx2U

[532] https://www.theguardian.com/science/2016/oct/30/neil-dregrasse-tyson-astrophysics-mars-exploration

[533] https://www.amnh.org/explore/science-bulletins/astro/documentaries/geologists-on-mars/why-go-to-mars/

[534] http://web.archive.org/web/20130102090721/https:/www.amnh.org/explore/science-bulletins/astro/documentaries/geologists-on-mars/why-go-to-mars/

[535] https://archive.org/download/DarkMissionTheSecretHistoryOfNASA_201611/Richard Hoagland/Dark Mission - The Secret History Of NASA.pdf

[536] http://www.wheredidthetowersgo.com/

[537] http://www.sacred-texts.com/ufo/hoagland.htm

[538] http://www.v-j-enterprises.com/smcnect.html

[539] http://www.ldolphin.org/LTDres.htmlhttp:/watch.pairsite.com/dolphin.html

[540] https://fas.org/irp/program/collect/stargate.htm

[541] https://www.newscientist.com/article/mg13318030-600-fatal-explosion-closes-cold-fusion-laboratory/

[542] https://mindcontrolblackassassins.com/tag/col-john-b-alexander/

[543] https://www.cia.gov/library/readingroom/document/cia-rdp96-00787r000500410001-3

[544] http://www.enterprisemission.com/moon1.htm

[545] http://www.reddirtreport.com/general/film-review-kubricks-odyssey-jay-weidner

[546] http://hartzog.org/j/2001sentinelanalysis.html

[547] https://saturn.jpl.nasa.gov/resources/73/

[548] http://web.archive.org/web/20070203113317/http:/www.comicbookresources.com:80/columns/oddball/index.cgi?date=2003-11-14

[549] http://thehiddenrecords.com/mars_1958_discovery.htm

[550] https://subterraneanpress.com/city-and-the-stars-and-against-the-fall-of-night-centpress

[551] http://www.isfdb.org/cgi-bin/pl.cgi?113181

[552] https://nthmind.wordpress.com/2012/02/16/comics-are-magic-part-2-using-superheroes-for-divination-and-manifestation/

[553] http://www.enterprisemission.com/moon3.htm

[554] https://photojournal.jpl.nasa.gov/catalog/PIA12739

[555] http://www.lulu.com/gb/en/shop/carl-james/science-fiction-and-the-hidden-global-agenda-2016-edition-volume-one/paperback/product-23209429.html

[556] http://www.imdb.com/title/tt0045917/

[557] http://www.imdb.com/title/tt0100802/

[558] http://www.imdb.com/title/tt0401729/

[559] http://www.imdb.com/title/tt0183523/

[560] http://brucedepalma.com/

[561] http://www.enterprisemission.com/m2m2.html

[562] https://www.youtube.com/watch?v=dgFN0U380vk

563 https://www.youtube.com/watch?v=xSY6on-M9Vw

564 https://www.space.com/8059-truth-photos-hubble-space-telescope-sees.html

565 http://marsrovers.jpl.nasa.gov/gallery/all/2/p/013/2P127520307EFF0309P2365L7M1.JPG

566 http://www.news.com.au/technology/conspiracy-theorists-confident-photoshopped-nasa-image-is-a-cover-up/news-story/b06c85cd52e9b92fa3c26902b222e8cf?sv=d703f8da08edf95871a037c6a86e3cb5

567 https://apod.nasa.gov/apod/fap/image/1004/titandione_cassini_big.png

568 https://apod.nasa.gov/apod/ap100420.html

569 http://www.gillevin.com/Mars/Reprint83-lifemars-files/Reprint83-lifemars.htm

570 https://link.springer.com/article/10.1007/BF01732380

571 https://apod.nasa.gov/apod/ap000626.html

572 http://www.abovetopsecret.com/forum/thread184836/pg1

Lightning Source UK Ltd.
Milton Keynes UK
UKHW020633180821
389043UK00006B/155

9 781981 117550